Inspection, Evaluation and Maintenance of Suspension Bridges

Inspection, Evaluation and Maintenance of Suspension Bridges

Edited by
Sreenivas Alampalli

New York State Department of Transportation

William J. Moreau

Chief Engineer, New York State Bridge Authority

CRC Press
Taylor & Francis Group
Boca Raton London New York

CRC Press is an imprint of the
Taylor & Francis Group, an **informa** business
A SPON PRESS BOOK

CRC Press
Taylor & Francis Group
6000 Broken Sound Parkway NW, Suite 300
Boca Raton, FL 33487-2742

First issued in paperback 2019

ISBN-13: 978-0-4665-9686-3 (hbk)
ISBN-13: 978-0-367-86852-9 (pbk)

Visit the Taylor & Francis Web site at
http://www.taylorandfrancis.com

and the CRC Press Web site at
http://www.crcpress.com

Inspection, Evaluation and Maintenance of Suspension Bridges

Edited by

Sreenivas Alampalli

New York State Department of Transportation

William J. Moreau

Chief Engineer, New York State Bridge Authority

CRC Press
Taylor & Francis Group
Boca Raton London New York

CRC Press is an imprint of the
Taylor & Francis Group, an **informa** business

A SPON PRESS BOOK

CRC Press
Taylor & Francis Group
6000 Broken Sound Parkway NW, Suite 300
Boca Raton, FL 33487-2742

First issued in paperback 2019

© 2016 by Taylor & Francis Group, LLC
CRC Press is an imprint of Taylor & Francis Group, an Informa business

No claim to original U.S. Government works

ISBN-13: 978-0-4665-9686-3 (hbk)
ISBN-13: 978-0-367-86852-9 (pbk)

**Visit the Taylor & Francis Web site at
http://www.taylorandfrancis.com**

**and the CRC Press Web site at
http://www.crcpress.com**

To all the bridge engineers around the world for
their dedicated service to public safety.

Sreenivas Alampalli and William J. Moreau

Contents

Preface

Owners and operators of long-span suspension bridges are scattered around the globe, just like the bridges they maintain. Some of the oldest long spans are in the United States; the newest are in and around the Pacific Rim; and Europe is home to many world-class suspension bridges, with many built post–World War II. New construction continues throughout Europe in the Netherlands, the United Kingdom, and Turkey. An array of interesting concepts to connect Italy with the island of Sicily are also on the drawing board.

One of the challenges for long-span suspension bridge operators is that most oversee only one or two bridges through small, single-purpose public authorities or subdivisions of state-run transportation agencies. This, combined with the fact that most suspension bridges operate over a very long lifetime, makes it difficult for bridge owners to acquire the experience necessary to detect signs of deterioration early, develop effective mitigation plans, and implement the appropriate restoration in a timely and cost-effective manner.

The International Cable Supported Bridge Operators' Association (ICSBOA) was conceived in 1991 when more than 125 international suspension bridge owners and operators assembled in Poughkeepsie, New York, to discuss common concerns, present research papers, and observe the main cables of the Mid-Hudson Bridge, which was undergoing a full-length main cable rehabilitation project at the time. Attendees traveled from Europe, South America, Asia, and across the United States to share problems, solutions, and best practices with the goal of reducing this challenge.

This book assembles decades of knowledge and experience through the authorship of many of the original attendees at the 1991 conference. Other progressive suspension bridge owners have joined the association and have also participated in creating this book. Subject matter is divided into discrete chapters, ranging from inspection and safety to main cable rehabilitation and strengthening. Each author offers his or her own perspective of the nuances of each bridge element unique to suspension bridges. In the end, the state of the practice, including a historical overview, design, inspection, evaluation, maintenance, and rehabilitation are covered in the book. A companion book will be published in the near future that illustrates the history of current operations of selective suspension bridges around the world in more detail.

Chapter 1 discusses, in chronological order, examples of engineering greatness achieved by numerous heroes of the suspension bridge industry. This chapter also examines the limited structural theory and material limitations of each era and how these factors affected early bridge designs and prompted experimentation, theorization, and discovery. Chapter 2 focuses on design and construction of suspension bridges, including various structural components that make up the suspension bridge, types of suspension bridges, general design philosophy and procedures, and construction sequence. Once the bridge is constructed, periodic inspections play a major role in ensuring bridge safety and maintaining the required level of service and long-term durability of the structure. Inspection results play a major role in determining appropriate short- and long-term cost-effective bridge management actions, including periodic and corrective maintenance actions, rehabilitation of various bridge components, and eventually, replacement of the major components or the entire structure. Chapter 3 covers visual inspection aspects of suspension bridges based on the authors' experience maintaining these bridges in the United States.

Evaluation of the bridge based on observed deterioration and condition of the bridge is a key aspect of maintaining any bridge. Given the complex nature of the suspension bridges and the cost associated with the maintenance and rehabilitation of these structures, evaluation is at best very involved and difficult in nature. Along with classical evaluation methods, probabilistic methods are required to analyze the main cables of the suspension bridges. These aspects, available methodologies to calculate remaining cable strength, and pros and cons of two evaluation tools are presented from a bridge owner's perspective in Chapter 4.

Corrosion of main cables is a key factor affecting the condition of the main cables in suspension bridges, as high-strength metallic wires are used in them and are exposed to corrosion-inducing weather conditions. These are very hard to inspect, as illustrated in previous chapters, and also are not easy to evaluate. Due to safety concerns and the high cost associated with any work on these suspension cables, understanding how corrosion affects these cables, how design and construction practices affect corrosion rates, and how the environmental effects drive deterioration mechanisms is a very important issue for current and future bridge owners. Chapter 5 discusses these aspects from a suspension bridge owner perspective based on years of experience maintaining them.

Suspension bridges are key components of our transportation network and often are critical to the network due to the traffic volumes they carry. Thus, maintenance plays a critical role in ensuring safety and the required level of service. Chapters 6 through 11 focus on maintenance aspects. Chapter 6 covers cable and suspender rope maintenance aspects. The authors took a

comprehensive approach and cover all aspects of these bridges, from fabrication and construction to maintenance of these components.

Cable dehumidification is emerging as a popular and effective corrosion mitigation strategy for old and new suspension bridges. Main cable dehumidification has been accomplished along the full length of the main cables in Japan, Denmark, and Scotland. Main cable preservation through dehumidification is the new direction for cable protection within the industry. Only time will tell if this strategy is successful as a single solution or if hybrid strategies are needed to achieve full protection. Chapter 7 comprehensively discusses dehumidification systems, based on the author's experience in maintaining suspension bridges in Japan. The chapter covers specification, system types, installation, how well they have been working, evaluation, and modifications.

Cable bands and their bolts are friction-critical connections requiring routine inspection and cyclical retensioning. The various techniques for controlled retensioning are presented in Chapter 8 with their respective advantages and limitations. Maintaining anchorages and their enhancements are a key aspect of suspension bridges. Chapter 9 identifies common problems, mitigation attempts, and their effectiveness associated with maintaining anchorage enhancements based on a survey of bridge owners. Chapter 10 discusses the maintenance aspects of anchorages based on the authors' experiences and knowledge of the methods and solutions used for a major historic bridge in the United States.

Preventive maintenance alone may not be enough in preserving the structural capacity in light of deterioration and ever-increasing traffic and other demands. Even though suspension bridge structures are unique and complex, many components and details used in these structures are not unique and are used in other structures. Thus, these can be rehabilitated, replaced, or strengthened to meet the demands through similar procedures used in other industries. But, replacing, rehabilitating, or strengthening unique elements such as cables can be challenging and require innovation and careful planning. Chapter 11 discusses such a case of adding to the factor of safety, as well as live load capacity, by widening the roadway and using a supplementary cable system.

Security and reducing time on target are imperative objectives in today's society. Protection of world-class suspension bridges now includes reducing exposure to the detrimental effects of terrorism. Successful strategies and risk-based assessments, including matrix analysis of counterterrorism enhancements, are covered in Chapter 12. Overall procedures and methodology are covered without going into details due to sensitivities associated with this aspect.

In essence, all aspects of suspension bridges—including history, design, inspection, maintenance, and rehabilitation—are thoroughly discussed based on the experience and knowledge of the owners maintaining these bridges for decades. Structural health monitoring can be an important strategy as suspension bridges age. Systematic and methodical record keeping is imperative to an effective monitoring system. Options available for structural monitoring using modern technology are presented in several chapters, along with the various expected results and limitations.

Owners and operators of suspension bridges are the current caretakers of these icons of the civil engineering profession. Modern technology, materials, and the lessons learned provide a great opportunity to improve the performance of, and extend the useful life of, these vital assets in transportation. Sharing these rare and technical experiences can only enhance our overall performance in this noble endeavor.

Sreenivas Alampalli and William J. Moreau
Editors

Acknowledgments

It has been a great pleasure working with the owners of suspension bridges around the world to bring this book to fruition. The discussions to document the body of knowledge gained by the current and past suspension bridge owners originated at a meeting of the International Cable Supported Bridge Operators' Association (ICSBOA) workshop held in May 2012 in New York State. It was decided by the editors that documenting these experiences in the form of two books will serve not only the owners, but also the entire bridge industry. This volume discusses the state of the practice in suspension bridge inspection, evaluation, and rehabilitation methods used worldwide. Its companion volume, which will be published in early 2016, discusses specific bridges around the world to give a comprehensive picture of how suspension bridges are maintained around the world. Knowing contributing authors do not have the time to undertake such an effort as this, but nonetheless spent valuable time documenting their experiences and practices for future generations of engineers, we thank them for their time and efforts in making this book—and the companion book that will be published soon—a reality.

We also thank our families (Sharada Alampalli and Sandeep Alampalli, and Cheryl Moreau) for their support during the preparation of this book.

Sreenivas Alampalli and William J. Moreau

Editors

Dr. Sreenivas Alampalli, PE, MBA, is director of the Structures Evaluation Services Bureau at the New York State Department of Transportation (NYSDOT). Before taking up the current responsibility in 2003, Dr. Alampalli was director of the Transportation Research and Development Bureau, where he worked for about 14 years in various positions. He also taught at Union College and Rensselaer Polytechnic Institute as an adjunct faculty.

Dr. Alampalli earned his PhD and MBA from Rensselaer Polytechnic Institute, his MS from the Indian Institute of Technology (IIT), Kharagpur, India, and his BS from Sri Venkateswara University, Tirupati, India. His interests include infrastructure management, innovative materials for infrastructure applications, nondestructive testing (NDT), structural health monitoring, and long-term bridge performance. Dr. Alampalli is a fellow of the American Society of Civil Engineers (ASCE), American Society for Nondestructive Testing (ASNT), and International Society for Health Monitoring of Intelligent Infrastructure (ISHMII). He was the recipient of the Bridge NDT Lifetime Service Award in 2014 from ASNT for outstanding voluntary service to the bridge and highway NDT industry. In 2013, he also received ASCE's Henry L. Michel Award for Industry Advancement of Research. Other notable awards he received include the ASCE Government Civil Engineer of the Year in 2014, and the prestigious Charles Pankow Award for Innovation from the Civil Engineering Research Foundation in 2000. He has authored or coauthored more than 250 technical publications, including two books on infrastructure health in civil engineering.

Dr. Alampalli is an active member of several technical committees in the Transportation Research Board (TRB), ASCE, and ASNT. He currently chairs the TRB Standing Technical Committee on Field Testing. He served as the TRB representative for NYSDOT and also as a member of the National Research Advisory Committee (RAC). He is a book review editor of the ASCE's *Journal of Bridge Engineering* and serves on the editorial boards of the journals *Structure and Infrastructure Engineering: Maintenance, Management, Life-Cycle Design and Performance* and *Bridge Structures: Assessment, Design and Construction.*

William J. Moreau, PE, served as the chief engineer of the New York State Bridge Authority for more than 27 years. Two of the six Hudson River crossings previously under his care were suspension bridges, constructed circa 1924 and 1930. Maintenance and preservation of these world-class suspension bridges became a career objective for him, culminating with service on the peer review panel for the development of National Cooperative Highway Research Program (NCHRP) Report 534: "Inspection and Strength Evaluation of Suspension Bridge Parallel-Wire Cables." Moreau participated for many years as a member of the Transportation Research Board committees and served as chairman of the Construction Committee for Bridges and Structures from 2005 to 2007.

Many early main cable inspection techniques, evaluation methods, and restoration materials were developed through partnerships Moreau developed with truly outstanding members of the engineering, construction, and material manufacturing communities in pursuit of arresting the effects of age and environment on the main cables of the Hudson Valley suspension bridges. He is currently semiretired and continues consulting in the New York City area.

Contributors

Sreenivas Alampalli
New York State Department of
  Transportation
Albany, New York

Mark Bulmer
AECOM
Leeds, United Kingdom

Yimin Chen
MTA Bridges and Tunnels
Metropolitan Transportation
  Authority
New York, New York

Danny K. Cobourne
Port Authority of New York/
  New Jersey
New York, New York

Charles Cocksedge
AECOM
London, United Kingdom

Barry Colford
Forth Road Bridge
Scotland, United Kingdom

Mohammed M. Ettouney
Weidlinger Associates, Inc.
New York, New York

Daniel G. Faust
AECOM
and
Delaware River Port Authority
Camden, New Jersey

Morys Guzman
Port Authority of New York/
  New Jersey
New York, New York

Lloyd Hansen
MTA Bridges and Tunnels
Metropolitan Transportation
  Authority
New York, New York

Russell Holcomb (retired)
New York City Department of
  Transportation
New York, New York

David List
Tamar Bridge and Torpoint Ferry
Plymouth, United Kingdom

Peter McDonagh
MTA Bridges and Tunnels
Metropolitan Transportation
  Authority
New York, New York

William J. Moreau
HAKS
New York, New York

Jaroslaw Myszczynski
MTA Bridges and Tunnels
Metropolitan Transportation
    Authority
New York, New York

Katsuya Ogihara
Honshu-Shikoku Bridge
    Expressway Co., Ltd. (HSBE)
Sakaide, Japan

Dora Paskova
MTA Bridges and Tunnels
Metropolitan Transportation
    Authority
New York, New York

Stewart Sloan
Port Authority of New York/
    New Jersey
New York, New York

Beverley Urbans
AECOM
Leeds, United Kingdom

Judson Wible
Parsons Brinckerhoff, Inc.
Ft. Lauderdale, Florida

David Wilkinson
AECOM
Leeds, United Kingdom

Bojidar Yanev
New York City Department of
    Transportation
New York, New York

Chapter 1

Suspension Bridges

An Overview

Bojidar Yanev

CONTENTS

1.1 FORM AND MATERIAL

All viable designs of human intelligence have natural analogs. In the fourth century B.C., Aristotle, learning from nature, identified the necessary ingredients in the life cycle of a successful structure as material, plan, execution, and service. Suspension bridges stretch all of these ingredients to their allowable limits. Appropriately, they were to take firm hold among the greatest structural achievements several millennia after their monumental predecessors made of timber, stone, and other naturally found materials.

Waddell (1916) illustrated the first volume of his *Bridge Engineering* with a timeless monkey bridge. The first suspension structures were natural fiber links. Nevertheless, suspension bridges could be engineered to Aristotle's specifications only under the new equilibrium of production and commerce introduced by the Industrial Revolution (Billington, 1983). Since that time, suspension bridges and their creators have consistently extended each other's physical reach and mental grasp. Functioning at the outer boundaries of performance and intelligence, the two must integrate seamlessly process and product, form and material, design and construction, theory and empiricism.

In his remarks on the comparative merits of cable and chain bridges (1841), John Roebling (1806–1869) cautions that "to ensure the successful introduction of cable bridges into the United States, their erection, and especially the construction of the first specimen, should not be left to mere mechanics." He seeks to integrate "the practical judgment" of "the most eminent Engineers" and the "rich store of scientific knowledge." Over the ensuing 172 years these two ingredients of the constructive process cooperated and competed in a "creative tension" (Billington, 1983, p. 52), producing the form and material of the longest tensile structures.

1.2 ART AND SCIENCE

Billington (1981) views the engineering of suspension bridges as a synthesis of art and science where form occasionally guided function. John Roebling and the Brooklyn Bridge serve as his primary examples. Kranakis (1997) analyzes the development of suspension bridges as a two-prong theoretical and empirical advance. Claude-Louis Navier (1785–1836) and his *Memoire sur les ponts suspendus* (1823) prominently represent the theoretical school. The bridges of John Finley (1762–1828), Thomas Telford (1757–1834), and Isambard Kingdom Brunel (1806–1859) exemplify the empirical approach.

The Maryland-born landowner and judge John Finlay obtained the first patent for a chain suspension bridge in 1808. Kranakis (1997, p. 43) describes his method as empirico-inductive, in the spirit of his contemporaneous thinkers of the Common Sense school. Thomas Paine (1737–1809) championed the new design. Finley modestly anticipated that "our puny canoe, with a little cultivation of genius, will soon spring into a formidable ship" (Kranakis, 1997, p. 53). The forecast proved prophetic. The chain suspension bridges designed by Telford and Brunel spanned 176 m at the Menai Straits (1826) and 214 m at Clifton (1864), respectively. In 1867, John Roebling commenced the construction of the Brooklyn Bridge between Manhattan and Brooklyn with spans of 284/487/284 m, supported by four parallel wire cables.

Metallurgy improved the material capable of reliably resisting high tension, from forged iron to high-strength steel. Design maximized the performance of this material by refining the suspension, the cable-stayed and various hybrid structural forms. Technological supply and transportation demand reshaped the main suspension elements from chains and ropes to helical strands, air-spun parallel wires, and prefabricated parallel wire strands. Construction grew capable of hoisting into position 4073 m long strands of 127 parallel wires and sinking 15,300-ton steel caissons 60 m below sea level. Theory advanced from empirical trial and error to nonlinear dynamic modeling, executable only by digital computers. By the end of the twentieth century, suspension spans reached the length of 1991 m. Another "puny canoe" in Finlay's creative stream was to launch the growing fleet of cable-stayed bridges.

Table 1.1 contains a partial list of noteworthy suspension bridges constructed since 1808 worldwide. Understanding the present and designing the future suspension bridges benefits from a review of several unique elements in their structure (i.e., the product) and stages in their life cycle (i.e., the process), as they have evolved over the encompassed period.

1.3 PROCESS AND PRODUCT

Engineering meets the physical demand of the applied loads with a supply of structural strength. Economy generates the financing for bridge

Table 1.1 Partial List of Noteworthy Suspension Bridges

Year	Bridge	Span Length (m)	Suspension	Stiffening	Towers
1808	Essex-Merrimack	40	2 IC	2 trusses	Masonry
1823	Saint Antoine	2 × 40	6 PIW	2 trusses	Masonry
1826	Menai Straits	120/176/120	EBC	2 trusses	Masonry
1834	Fribourg	273	4 PW	2 trusses	Masonry
1849	Wheeling (1872)	308	4 PW/SR	2 trusses	Masonry
1864	Clifton	214	EBC	2 trusses	Masonry
1866	Cincinnati-Covington	322	4 AS/SR	2 trusses	Masonry (Figure 1.1)
1883	Brooklyn	284/487/284	4 AS/SR	6 trusses	Masonry (Figure 1.2)
1903	Williamsburg	488	4 AS[b]	2 trusses	Steel (Figure 1.2)
1909	Manhattan	222/449/222	4 AS	4 trusses	Steel (Figure 1.2)
1926	Bear Mountain	688	2 AS	2 trusses	Steel
1928	Point Pleasant	213.5	2 EBC	2 trusses	Steel
1931	George Washington	186/1067/198	4 AS	4 trusses	Steel (Figure 1.3)
1962	Lower Leaf	186/1067/198	4 AS	Space truss	Steel (Figure 1.3)
1937	Golden Gate	343/1280/343	2 AS	Space truss	Steel (Figure 1.3)
1939	Bronx-Whitestone	224/701/224	2 AS	2 girders	Steel
1940	Tacoma Narrows	396/792/396	2 AS	2 girders	Steel
1957	Mackinac	549/1158/549	2 AS	Space truss	Steel
1964	Verrazano	371/1300/371	4 AS	Space truss	Steel (Figure 1.3)

(continued)

Table 1.1 (continued) Partial List of Noteworthy Suspension Bridges

Year	Bridge	Span Length (m)	Suspension	Stiffening	Towers
1981	Humber	280/1410/530	2 AS	Box girder	R/C concrete
1988	Kita Bisan-Seto	274/990/274	2 PWS	Space truss	Steel
	Minami Bisan-Seto	274/1100/274	2 PWS	Space truss	Steel
1995	Askøy	850	2 × 21 LCS	Box girder	Concrete
1996	Jiangyn	1385	2 AS	Box girder	Concrete
1996	Akashi Kaikyo	960/1991/960	2 PWS	Space truss	Steel (Figure 1.4)
1997	Tsing Ma	356/1377	2 AS	Box girder	Concrete (Figure 1.5)
1998	Great Belt	535/1624/535	2 AS	Box girder	Concrete (Figure 1.5)
1999	Kurushima I	149/600/170	2 PPWS	Box girder	Steel (Figure 1.6)
	Kurushima II	1020/250	2 PPWS	Box girder	
	Kurushima III	1030	2 PPWS	Box girder	
2012	Taizhou	1080/1080	2	Box girder	Central—steel 2 sides—concrete
2013	East Bay Crossing	385	1 PPWS[a]	Box girder	Steel (Figure 1.7)
2016	Izmit Bay	566/1550/566	2	Box girder	Steel
	Stretto di Messina	183/3300/183	4	3 linked box girders	

Note: AS, aerially spun parallel wires; EBC, eyebar chains; HWS, helical wire strands; IC, iron chains; LCS, locked-coil strands; PIW, parallel iron wires; PPWS, parallel wire strands; PWS, parallel wire strand; SR, steel rope stays.

[a] Self-anchored.

[b] Nongalvanized.

Figure 1.1 The Cincinnati-Covington Bridge (1866).

Figure 1.2 The Brooklyn (1883), Williamsburg (1903), and Manhattan (1909) Bridges, New York City.

construction and operation from the current and anticipated demand for transportation. During the reviewed period, that demand evolved from horse-drawn carriages to trains, then predominantly to internal combustion vehicles, and may be moving on to new forms of rail transport. In 1855, Roebling spanned 251 m at Niagara Falls with a "double decker," carrying trains on top and carriages below. The most ambitious spans of the early twentieth century emphasized rail transport and readily accommodated vehicular traffic. The Williamsburg Bridge (1903) carries eight

Figure 1.3 The George Washington (1931), Verrazano (1964), New York, and Golden Gate (1937), San Francisco, Bridges.

Figure 1.4 The Akashi Kaikyo (1996) Bridge.

vehicular lanes and two rail tracks. Manhattan Bridge (1909) carries seven traffic lanes and four rail tracks. By midcentury, vehicular traffic volume dictated that the George Washington Bridge would carry 14 (1962) and the Verrazano Bridge 12 traffic lanes (1964). The relatively lighter, purely vehicular traffic encouraged the growth of the longest spans. Mixed-mode service has not entirely disappeared, however. The Bisan-Seto and Tsing Ma Bridges carry six traffic lanes and two rail tracks. The Third Bosphorus Bridge is built for both vehicles and trains. Mixed-mode use is proposed for the Messina Straights Bridge.

Figure 1.5 The Tsing Ma (1997), Hong Kong, and Great Belt (1998), Denmark, Bridges.

Figure 1.6 The West Bay (1937), San Francisco–Oakland, and Kurushima (1999), Japan, Bridges.

The engineering product evolving under these dynamic social and physical constraints must maximize the efficient use of the strongest, lightest, and toughest structural materials. The resulting structure is the suspension bridge, with its unique cables, suspenders, towers, anchorages, saddles, decks, and stiffening systems. Equally unique is the associated process, including the analysis, design, construction, maintenance, repair, and replacement of these elements.

Figure 1.7 The East Bay San Francisco–Oakland Bridge during construction (2011).

1.4 SUSPENSION

1.4.1 Wires versus Chains

Chains of one kind or another supported most of the iconic early suspension bridges. The chain material evolved from forged wrought-iron links (1796) to wrought-iron eyebars (1818) and steel eyebars (1828). After 1915, nickel and heat-treated carbon steel alloys increased the ultimate strength of the eyebars to 105 ksi (75.6 kg/mm², 741 MPa). By the year 2000, electric arc furnaces had replaced the traditional open-hearth Siemens–Martin process.

Under the title "Wire Cables vs. Eyebar Chains" (1949, p. 74), David Steinman (1887–1960) allows that for span lengths of up to 214 m (700 ft), as in the bridges over the Ohio River at Point Pleasant and St. Mary's, heat-treated eyebars may be cost-competitive. Their principal advantages are the superior stiffness, speedier construction, and simpler global structural behavior—and hence analysis. The ultimate choice is reduced to the following reasoning: "Where two designs are of equal cost, the heavier bridge is to be preferred as giving a more rigid structure" (Steinman, 1949, p. 756). By the first writing of this text in 1922, John Alexander Low Waddell (1854–1938) was still championing trusses as the best long-span bridges. His foremost detractor, Gustav Lindenthal (1850–1935), was trying to graft eyebar trusses onto suspension spans of record length.

Against eyebars, Steinman cites high secondary and unequally distributed stresses, difficulty of inspection and painting, and "untried problems in the erection." The collapse of the bridge at Point Pleasant, 100 years after the commencement of the Brooklyn Bridge construction, settled the debate. The twin bridge at St. Mary's was promptly decommissioned.

Multiple eyebar chains have remained in service without incident, for example, on the cantilever trusses of the Queensboro Bridge over East River (Figure 1.8b) in New York City and the three self-anchored suspension bridges over the Allegheny River (Figure 1.8c) in Pittsburgh. The eyebar chains of historic suspension bridges, such as the Széchenyi Bridge over the Danube in Budapest (Figure 1.8a) and Telford's bridge over the Menai Straits, have been reconstructed reproducing as much as possible the original form, if not the material.

The steel wires of modern cables are strengthened by successive drawing (as opposed to extrusion) through dies with decreasing diameters. Cold drawing modifies the molecular configuration of mild steel, increasing considerably its yield and ultimate strength. The elastic modulus of the steel is retained; however, the ductility is reduced. Upon attaining their final diameter (4.8–7.0 mm), the wires are "hot dipped" in molten zinc. A zinc coating with a thickness of 20–40 μm forms over the steel surface, and provides galvanic protection against corrosion. Figure 1.9 shows different wires used in suspension bridge parallel wire cables. Figure 1.10 shows the "wire certificate" for the air-spun cables of the Great Belt Bridge.

In a noteworthy exception, the four cables of the Williamsburg Bridge consist of 7696 nongalvanized high-strength parallel wires. The cables were extensively rehabilitated in the 1990s. Also nongalvanized were the original helical strand cables of the Pont de Tancarville (176/608/176 m) over the Seine and Pont d'Aquitaine (143/394/143 m) over the Garonne in France (Figure 1.11). The cables at both bridges were replaced, in 1998 and 2002, respectively.

Heat treatment and refinements in the chemical composition of the alloy consistently increased the tensile strength of cable wires. A wire strength of 1600 MPa (155–160 kg/mm^2) was prevalent throughout most of the twentieth century, as reported by Gimsing (1997) and presented in Table 1.2.

Figure 1.8 Széchenyi (1849, 1949), Budapest; Queensboro (1909), New York; and 6th St. Bridge (1928), Pittsburgh.

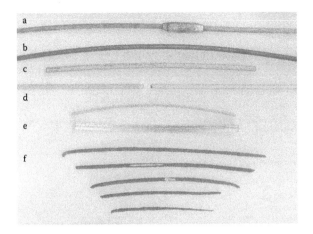

Figure 1.9 High-strength wires used in suspension bridge cables: (a) 4.8 mm diameter, galvanized, with splicing ferrule, Manhattan Bridge, 1909; (b) 4.8 mm diameter, nongalvanized, Williamsburg Bridge, 1903; (c) 5.37 mm diameter, galvanized, Great Belt Bridge, 1996; (d) 4.8 mm diameter, galvanized, heat straightened, ductile break; (e) Z-shaped galvanized wrapping wire, Kurushima Bridge, 1999; and (f) nongalvanized, corroded wires, Williamsburg Bridge, 1988.

Mayrbaurl (2006) reported strengths from 1644 to 1695 MPa for wires from three suspension bridges identified as W, X, and Z. According to Nishino et al. (1994), the record span of the Akashi Kaikyo Bridge required the higher strength of 1975 MPa (180 kg/mm²), as shown in Figure 1.12.

1.4.2 Strands: Parallel Air-Spun, Prefabricated, and Helical

Table 1.2 contains a comparison of the mechanical properties (Gimsing, 1997) and some performance attributes of the different strands. The following three methods for assembling strands, and ultimately cables, from high-strength steel wires have evolved into current coexistence:

- Parallel wires by air spinning (AS) (Figure 1.13a)
- Prefabricated parallel wire strands (PPWSs) (Figures 1.13b, 1.14, and 1.15)
- Helical wire strands (HWSs) (Figure 1.14)

Mayrbaurl and Camo (2004) reported 52 suspension bridges in North America, of which 29 have aerially spun parallel wire cables (AS), 21 have HWSs, and 2 have PPWSs.

John Roebling introduced the air-spinning method of cable construction around 1845 as an application of his high-strength wires. The Roebling Wire Company fabricated high-strength galvanized wires and steel ropes

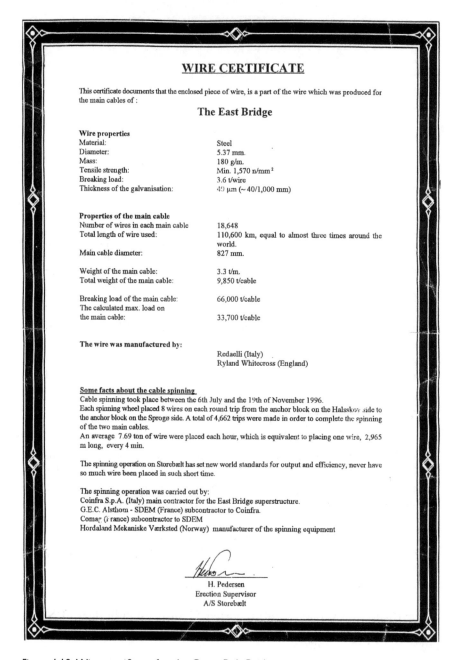

WIRE CERTIFICATE

This certificate documents that the enclosed piece of wire, is a part of the wire which was produced for the main cables of :

The East Bridge

Wire properties

Material:	Steel
Diameter:	5.37 mm.
Mass:	180 g/m.
Tensile strength:	Min. 1,570 n/mm^2
Breaking load:	3.6 t/wire
Thickness of the galvanisation:	40 µm (~ 40/1,000 mm)

Properties of the main cable

Number of wires in each main cable	18,648
Total length of wire used:	110,600 km, equal to almost three times around the world.
Main cable diameter:	827 mm.
Weight of the main cable:	3.3 t/m.
Total weight of the main cable:	9,850 t/cable
Breaking load of the main cable:	66,000 t/cable
The calculated max. load on the main cable:	33,700 t/cable

The wire was manufactured by:

Redaelli (Italy)
Ryland Whitecross (England)

Some facts about the cable spinning

Cable spinning took place between the 6th July and the 19th of November 1996.
Each spinning wheel placed 8 wires on each round trip from the anchor block on the Halsskov side to the anchor block on the Sprogø side. A total of 4,662 trips were made in order to complete the spinning of the two main cables.
An average 7.69 ton of wire were placed each hour, which is equivalent to placing one wire, 2,965 m long, every 4 min.

The spinning operation on Storebælt has set new world standards for output and efficiency, never have so much wire been placed in such short time.

The spinning operation was carried out by:
Coinfra S.p.A. (Italy) main contractor for the East Bridge superstructure.
G.E.C. Alsthom - SDEM (France) subcontractor to Coinfra.
Comag (France) subcontractor to SDEM
Hordaland Mekaniske Værksted (Norway) manufacturer of the spinning equipment

H. Pedersen
Erection Supervisor
A/S Storebælt

Figure 1.10 Wire certificate for the Great Belt Bridge.

Figure 1.11 Pont de Tancarville (1998) and Pont d'Aquitaine (1996), France.

for the cables and suspenders of all the record-breaking bridges built over the ensuing century.

The AS method consists of unreeling several wires from one anchorage, over the tower saddles, to the opposite anchorage. At the anchorages the wires go over a strand shoe and reverse direction. At their ends, wires are spliced with crimped ferrules, as in Figure 1.9a, or with threaded collars (considered detrimental to the strength of the connection). Steinman (1949) reports wire lengths of up to 1000 m (later exceeded).

Air-spun strands typically comprise 200 to 300 wires bundled and anchored together. Circular cylinders with the same diameter can be compacted around a central one in layers containing increasing multiples of 6, as in $7 = 1 + 6$, $19 = 1 + 6 + 12$, $37 = 1 + 6 + 12 + 18$, $61 = 1 + 6 + 12 + 18 + 24$, and so on.

Hexagonal configurations of a 7-wire rope and a prefabricated strand (PPWS) consisting of 127 wires are shown in Figure 1.14.

The air spinning of the two parallel wire cables at the Golden Gate Bridge introduced several innovations (Strauss, 1938). In order to reduce the lateral load on the strands during compaction over the saddles, the strands were aligned vertically, along the $X'OY'$ axes in Figure 1.16, rather than horizontally, along the traditional XOY axes. To better approximate a circular cross section (Figure 1.13), the 27,572 wires in each of the two cables were bundled in 122 strands of unequal size. The circular cross section of the Akashi Kaikyo cables was also achieved by adding wires to the polygonal shape obtained by compacting the 127-wire PPWSs.

While Roebling's AS cables remained prevalent on the American continent, European engineers retained a preference for steel wire ropes. With the decisive influence of Eugene Freyssinet (1879–1962), the practice evolved to prestressing strands (HWS). Multiwire helical strands are built of layers spun in alternating directions. Gimsing (1997) reports a nominal elastic

Table 1.2 Properties and Attributes of Modern High-Strength Strands with the Following Wire Properties: Diameter = 4.8–5.37 mm, Elastic Modulus E = 205 GPa, σ_{ult} = 1570 MPa, σ_y = 1180 MPa

	7-Wire	Parallel (AS)	Helical (HWS)	Locked-Coil	Prefabricated (PPWS)
Elastic modulus E (GPa)	190	205	170	180	200
Ultimate strength σ_{ult} (MPa)	1770–1860 (>1570)	1570	0.9×1570	1370–1570 (of individual Z-shaped wires)	1570
Positive		Familiar construction, accessible for inspection	Compaction, accelerated construction, maximized clamping effect	Compaction, accelerated construction, maximized clamping effect	Compaction, accelerated construction, improved clamping effect, reduced bending
Negative		Slower construction, reduced compaction, bending stresses	Limited span length, interior not inspectable, possible fretting	Limited span length, interior not inspectable, fatigue, fretting	Weight of strands may become prohibitive during construction

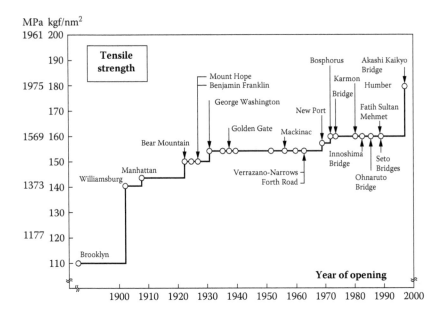

Figure 1.12 Evolution of tensile strength in cable wires from the Brooklyn to the Akashi Kaikyo Bridge.

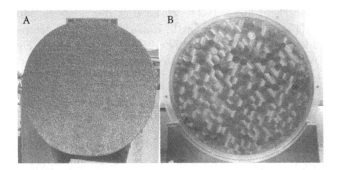

Figure 1.13 Mock-up cross sections of suspension bridge cables: (a) Golden Gate, 924 mm diameter, 27,572 wires, 122 air-spun strands, and (b) Akashi Kaikyo, 1120 mm diameter, 36,830 wires, 290 prefabricated strands.

modulus of such strands as roughly 15–25% lower than that of straight wires, reflecting the compaction of the strands under tension.

Steinman and Watson (1957, pp. 340–342) describe the introduction of "rope-strand cables" in America as follows:

This new type of suspension bridge construction has been introduced and developed by Robinson and Steinman, commencing with the Grand'Mere Bridge over the St. Maurice River in Quebec, completed in

1929 with main span of 950 feet (276 m); and it has since been used by them in the Waldo-Hancock Bridge, the St. Johns Bridge, the Thousand Islands Bridge, the Wabash River Bridge, the Deer Island Bridge, and the Lion's Gate Bridge at Vancouver.... The rope strand type of cable construction offers distinct economy, in saving time and labor in the field, for spans up to about 1,500 feet (457.5 m). For longer spans parallel wire cables remain the most economical form of construction.

Locked-coil strands are helical with outer layers made of Z-shaped wires, intended to protect the interior wires from the intrusion of humidity. They were used in the original cables at Pont de Tancarville and Pont d'Aquitaine (Figure 1.11), to cover internally nongalvanized strands.

Helical strands can have internal protective layers. With adequate external corrosion protection, including dehumidification, helical strands have shown no visible deterioration, for example, at the Little Belt Bridge (main span 600 m, 1970) (Figure 1.14). Fretting and stress concentration, particularly for the super-compacted locked-coil strands, are potential hazards. Helical strands remain the prevalent option for cable-stayed bridges; however, parallel wire strands similar to the suspenders of the Akashi Kaikyo Bridge (Figure 1.15) were used as stays at the record-breaking Tatara Bridge (890 m, 1999).

Prefabricated parallel wire strands were proposed by Jackson Durkee (1966) at Bethlehem Steel. He reported that if a 37-wire parallel wire strand is slightly twisted (or pitched), it can be reeled on drums of larger diameters, akin to helical strands. The PPWSs of 127 wires (Figures 1.14 and 1.15c) were used first in Japan and later worldwide. The two cables

Figure 1.14 (a) High-strength nongalvanized 7-wire steel rope. (b) Helical strand, Little Belt Bridge. (c) Prefabricated parallel 127-wire strands.

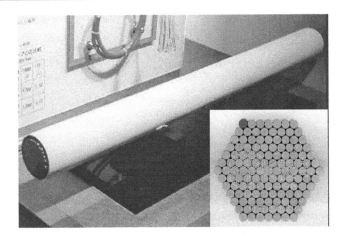

Figure 1.15 Prefabricated strand of 127 parallel wires, used in the Akashi Kaikyo cables and suspenders, and in the stays of Tatara Bridge.

of the Akashi Kaikyo Bridge (Figure 1.13b) are built of 290 PPWSs with a length of 4100 m.

1.4.3 Compaction

Cables are compacted as much as possible both as a protection against the corrosive intrusion of water and as a means of maximizing the clamping effect of adjacent wires. Perfectly compacted wires (except external ones) are in contact with six adjacent wires, and form hexagons, as shown in Figures 1.14 through 1.16.

For T perfectly compacted concentric layers of wires, as in Figure 1.16, the net-to-gross ratio of areas can be approximated as shown in Equation 1.1:

$$2\pi r^2 [3T(T + 1) + 1]/r^2 [(2T + 1)^2 3^{3/2}] \approx (2\pi/3^{3/2}) (3/4) \approx 0.907 \qquad (1.1)$$

The equilateral triangle defined by the centers of three identical tangent circles is the simplest repetitive unit of the hexagon and directly obtains the same result, as in Equation 1.2:

$$\pi r^2/(2r^2 3^{1/2}) = 0.907 \qquad (1.2)$$

This maximum compaction is reported for Z-shaped strands (Table 1.2). It is approximated for the heat-straightened strands shown in Figure 1.14c. The voids in aerially spun cables comprise normally about 20% of the gross section area. The compaction should be checked during inspections by comparing the cable's circumference and the number of wires in it. Figure 1.17

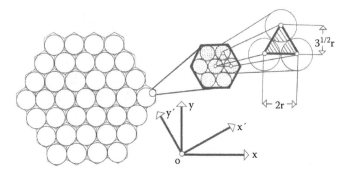

Figure 1.16 Compacted geometry of 37, 6, and 3 circles with radius *r*.

Figure 1.17 Parallel wire AS cable during wedging and compaction.

shows parallel wire AS cables wedged for inspection and compacted for rewrapping with spiral wire.

1.4.4 Parallel Wire Stress–Strain State

Traditionally, the product of the wire strength and the number of wires in a suspension cable was considered its capacity. The ratio of the capacity and the design load acting on the cable was perceived as a safety factor. The implied assumptions of uniaxial tension and linear elastic response obscure the following essential features of suspension cables:

- The wire stress state is complicated by residual stresses induced during galvanization.
- Stress concentration and embrittlement are caused by the presence of incipient cracks and hydrogen.

• During construction and service, the wires are subjected to flexure both by the straightening of their initial curvature and by additional flexure at saddles, strand shoes, and cable bands.

1.4.5 Wire Curvature

The fully extended galvanized wires cool off into a naturally curved shape with a diameter of approximately 2 m. Consequently, compaction into a cable subjects them to bending. Straightening a wire with a section radius r_{wire} and a curvature with radius R induces in it the following bending moment M and corresponding maximum stress σ:

$$M = E\ I/R \tag{1.3}$$

$$\sigma = M/S = E\ r_{wire}/R \tag{1.4}$$

where:
  $I = \pi\ r_{wire}{}^4/4$ is the moment of inertia of the wire section
  $S = I/r_{wire}$ is the wire section modulus

Introducing $r_{wire} = 5$ mm and $R = 2000$ mm in Equation 1.4 obtains $\sigma = 500$ MPa (75 ksi), matching the uniform working stress level for which many suspension cables have been designed. Based on x-ray diffraction tests, Mayrbaurl and Camo (2004) estimated that the straightening can induce bending stresses of up to ±240 MPa (36 ksi). Wires invariably crack on the side where straightening has produced tension.

1.4.6 Cable Bands, Saddles, and Anchorages

Gimsing (1997) obtains the local bending at the first cable band past the saddle as a function of the cable tension, diameter, change in the angle over the distance to the saddle, band length, assuming 20% voids. The local bending rapidly declines toward midspan.

To some degree air-spun wires can adjust their own curvature to that of the saddles during construction.

Despite their limited ductility, AS cable wires appear capable of sustaining the plastic deformation caused by wrapping around strand shoes in anchorages, as in Figure 1.18a. At the back of the strand shoe the wires are maximally deformed, as well as compacted, and experience no other load. Prefabricated and helical wire strands are not bent in the anchorages (Figure 1.18b and c); however, their sockets may be vulnerable in other ways.

Figure 1.18 Strand anchorages: (a) parallel air-spun, (b) helical, and (c) prefabricated.

1.4.7 Stress Distribution within the Parallel Wire Cable

The stress is likely to vary between cable wires and strands in a cable to the extent that their initial geometry cannot be perfectly identical. The Honshu-Shikoku Bridge Authority (1998) reports nightly adjustments of the prefabricated strands during the installation to ensure easier compaction.

With the aging of AS cables, it has become necessary (as it did at the Williamsburg Bridge in 1988) to evaluate their remaining strength after a number of wires have broken. It is generally accepted that over a certain development length, a broken wire will regain the stress–strain state of the adjacent wires through friction; however, that length cannot be rigorously quantified under an imperfect compaction and a varied level of stress. Starting with the model shown in Figure 1.16, Gjelsvik (1991) demonstrated that, under perfect compaction, the clamping length can be "typically a few feet," because the broken unstressed wire expands. Under imperfect field conditions, that length rapidly increases. For practical purposes, it has been assumed that three cable bands represent a sufficient development or clamping length for a single wire.

Mayrbaurl and Camo (2004) propose a variety of models for assessing the remaining strength of a deteriorated cable, depending on the available data for the strength, ductility, corrosion, and incipient cracking of tested wires. Since none of that information can be perfectly available, any assessment of the cable's strength must be a probabilistic one. The number of field observations and laboratory tests required for a meaningful analysis increases with the level of deterioration, and with the decrease of the design safety factor.

1.4.8 Cable Stiffness

The effective modulus of a suspended cable E_{ef} can be approximated by Equation 1.5, attributed to Ernst, as that of two springs in series, one with the elastic modulus of steel E, and the other, resulting from the sag, as follows:

$$1/E_{ef} = 1/E + (\gamma \ell)^2/(12\ \sigma^3) \tag{1.5}$$

where:

    ℓ is the span length
    γ is the specific weight of the cable
    σ is the stress in the cable

Since E_{ef} and σ are mutually dependent, an initial assumption and some iterations are required in every case. The effective stiffness E_{ef} decreases with ℓ^2, but increases with σ^3. The stress σ in turn increases with the span length ℓ until the cable can sustain only its weight. Thus, the estimated ratio of the cable strength to the maximum loads for the longest suspension cables has decreased from a factor of 4 at the beginning of the twentieth century to 2.2. That trend implies an increased intolerance to any form of deterioration. Gimsing (1997) refers to E_{ef} as E_{tan} and recommends the use of the secant modulus for cable-stayed bridges.

1.4.9 Number of Cables

John Finley concluded that four chains work better than two. The four cables at each of the three East River suspension bridges in New York City (Figure 1.2) are grouped in different configurations. The inclined cables on the Williamsburg Bridge recall a recommendation made by Roebling (1841). In the same article, Roebling observed that "a single cable will be superior to a pair of cables when displaced by lateral forces." The demands for superior stiffness, aerodynamic stability, and constructability have contributed to the trend of using two cables instead of four, as evidenced in Table 1.1. At the Messina Straits Bridge, which is designed for both vehicles and trains on a single level, four cables were proposed once again.

1.4.10 Suspenders

The evolution of suspenders has followed that of the main cables, resulting in three main alternatives: galvanized wire ropes, helical strands (Figure 1.19), and parallel wire strands (Figure 1.15).

Figure 1.19c shows a combination of wire ropes and strands. Solid rods have a limited use over shorter lengths, as, for example, at the Chelsea Bridge (Figure 1.20). On the Brooklyn Bridge short rods serve as both suspenders and compression struts.

On older bridges, such as the East River crossings shown in Figures 1.2 and 1.25, suspenders are spaced at 6 m. The PWS suspenders at the Akashi Kaikyo Bridge are spaced at 14.2 m and reach a length of 205 m.

Where the stiffening system of the bridge is a truss, the suspenders can be anchored at the top chord (as in more recent bridges) or the bottom

Figure 1.19 Suspenders: (a) wire rope, (b) wire ropes with spacers, and (c) wire ropes and prestressing helical strand.

Figure 1.20 Chelsea Bridge, London.

chord (as on older ones). Mixed anchoring can also become necessary due to geometric constraints.

Wire rope suspenders typically stride over grooved cable bands, as in the examples of Figures 1.3c and 1.19a and b. In the hybrid system shown in Figure 1.2a, the rope suspenders are socketed under the cable bands, in order to form a single vertical plane coincident with the plane of the diagonal stays fanning out from the top of the towers. Helical and parallel wire strands are socketed and pin connected, under equalizer bars, as in Figure 1.19c, or under the cable bands, as in the case of Figure 1.4b. In the latter example, the suspenders are connected with simple or universal pin joints, depending on their length.

$$1/E_{ef} = 1/E + (\gamma \ell)^{2}/(12 \, \sigma^{3}) \tag{1.5}$$

Suspenders perpendicular to the longitudinal stiffening system do not contribute to its stiffness. Freeman Fox & Partners designed inclined suspension systems for the Severn (305/988/305 m, 1966), the First Bosphorus (1074 m, 1973), and the Humber (280/1410/530 m, 1981) Bridges. Such suspenders can transmit shear between the main cables and the deck, thus dampening structural vibrations. As a result, they experience higher stress cycles of their own. Nets of intersecting suspenders inclined in opposite directions have also been proposed. Their action within the suspension system is comparable to that of tension diagonals in a truss, suggesting the name *cable trusses*. The five-span San Marcos Bridge in El Salvador (76/159/204/159/76 m) was an example of this system from 1951 to its demolition in 1981.

In the proximity of their sockets, suspenders experience highly variable stresses due to live loads and wind-induced vibrations, as well as the most aggressive corrosion. Consequently, design must anticipate their periodic replacement. To ensure continuous service, the suspenders and the longitudinal stiffening must be designed to redistribute the added load of a predetermined number of missing suspenders without irreversible global deformations. The suspenders on the East River bridges shown in Figure 1.2 have been replaced more than once, under traffic, for example, as shown in Figure 1.9c.

The dynamic characteristics of vertical suspenders running in groups of two or four can be adjusted by connecting them at selected points along their length with spacers, as in Figure 1.19b. The combination of wind and rainwater running down the polyethylene tubes encasing suspenders and stays has caused "galloping vibrations." The smooth surfaces of modern suspenders and stays are corrected in order to guide water flow. Dampers are installed at the bottom anchorages.

1.4.11 Corrosion Protection

Main cables are protected from corrosion by various systems. The galvanization of the wires is the first level of protection. Synthetic and natural corrosion inhibitors (such as linseed oil) are used to fill the voids. Wires have been coated with lead and, more recently, zinc paste. Parallel wire cables are wrapped with spiral wire, polyethylene sheets, and painted.

Many strand cables have no overall wrapping. Individual strands may have protective coating. It is assumed that tension compacts strands sufficiently to prevent water penetration, whereas a wrapping would trap moisture. Examples are the Pont de Tancarville and Pont d'Aquitaine (Figure 1.11) and the Chelsea Bridge (56/101/56, 1937), shown in Figure 1.20.

Suspenders made of ropes, prestressing strands, and parallel wire strands are protected differently. All are galvanized. Ropes are painted. Some helical strands rely solely on galvanization. Polyethylene tubes encase PPWSs, as in Figure 1.15.

1.5 TOWERS (PYLONS)

Gourmelon and Brignon (1989) define suspension bridge towers as the pedestals of the cables. Early towers consisted of individual pylons, as in the case of Roebling's bridge at Niagara Falls and the Chelsea Bridge (56/101/56 m, 1937) in London (Figure 1.20).

Roebling eventually opted for massive portal frames, as at his Cincinnati-Covington (Figure 1.1) and the Brooklyn (Figure 1.2) Bridges. Tower material evolved from wood to unreinforced concrete masonry (Figures 1.1 and 1.2a), steel (Figures 1.2, 1.3, and 1.6), and reinforced concrete (Figures 1.4 and 1.11) with posttensioning. Steel is the material of most American and Japanese bridge towers, whereas concrete is common in European ones. Earlier towers were designed as rigid with sliding saddles. Contemporary steel and reinforced concrete towers are designed to resist the lateral loads transmitted by fixed saddles in flexure. The 210 m towers of the Golden Gate Bridge were designed for movements of "18 in [460 mm] channelward or 22 in [560 mm] shoreward from vertical lines through the centers of the tower bases" (Strauss, 1938).

Towers of small bridges, for example, pedestrian ones, can be inclined and articulated at the base.

Whereas bridge users focus on the fleeting road, two geographically separate communities share similar monumental towers. The height of suspension bridge towers is roughly 10% of the span length; however, that amounts to 254.1 m at the Great Belt and 282.8 m at the Akashi Kaikyo Bridge. Thus, towers and the catenary shape form the trademark images of the suspension bridges. In 1867, Roebling reported to the New York Bridge Company (Reier, 1977) that "the most conspicuous features, the towers" of the proposed Brooklyn Bridge "will serve as landmarks to the adjoining cities, and they will be entitled to be ranked as national monuments." One hundred and fifty-seven years later, that ranking is worldwide. The 184 m tall towers of the George Washington Bridge have been praised for the "honesty" of their (unintended) functional look by critics as demanding as Le Corbusier. In contrast, the brutally utilitarian clunky towers of the Williamsburg Bridge and the deliberately ornamental slender ones of the Manhattan Bridge had to deliver indispensable service for a century in order to gain recognition. Some appearances have benefited from professional help. Irving F. Murrow, consulting architect, designed the striking Art Deco towers and the trademark "international orange" color of the Golden Gate Bridge. Gimsing (1998) states: "The concrete pylons are the most spectacular elements of the East Bridge and the development of their appearance through close collaboration between architects and engineers was a key issue in the designing process."

1.6 SADDLES

If towers are the cables' pedestals, saddles are their bearings. Saddle cur-
vatures are, for example, 7 m at the Great Belt and 9.15 m at the George
Washington Bridge. The rotated arrangement of cable strands from horizontal to vertical, shown in Figure 1.16, led to a modification from flat to
U-shaped (at the Golden Gate Bridge), and ultimately to "grooved" saddles
for prefabricated strands, as shown in Figure 1.21.

Saddles designed to slide tend to "freeze" in fixed positions, over time,
and the lateral forces transmitted by the cables to the tower tops increase.
During the repair of the saddle rollers at the Williamsburg Bridge in the
1990s, the previously frozen saddles shifted abruptly by almost 150 mm.
Fixed cable saddles were designed for the adjacent Manhattan Bridge
(1909) (Figures 1.2 and 1.25) and have become prevalent. The 180-ton
saddles on the George Washington Bridge were installed on beds of 41 steel

Figure 1.21 Saddles: (a) flat, (b) grooved, and (c) splay saddle.

rollers with 200 mm diameter and fixed after the addition of the lower deck in 1962. The saddles of the Golden Gate Bridge were similarly set on a nest of 200 mm rollers and permanently fixed after the completion of the construction.

If there is a large difference between the cable forces in the main and side spans, the tendency of the towers to lean may be counteracted by strands added to the "backstay" part of the cables. Such strands are typically anchored on top of the saddles. This is the case, for example, at the Tsing Ma Bridge (Figure 1.5), where one side span is not suspended, but also at the Mackinac Bridge, where the ratio of the side to main span is as high as 0.48. Scott (2001) reported that 240 wires were added to the 12,580-wire main cables in the backstays of the latter bridge.

1.7 ANCHORAGES

As suspended spans grow longer, their anchorages gain weight and rigidity. Where the terrain allows, even at record-breaking suspension bridges such as the Bear Mountain and George Washington, anchorages have been drilled into rock. The typical gravity anchorages rely on the weight of massive concrete anchor blocks (Figure 1.22). Key elements of the anchorages include the steel anchor girders, the cable anchor frames, the bent blocks, struts, and splay saddles.

Restrictions on the construction of large anchorages provide an incentive toward longer cable-stayed bridges, such as the Russky in Vladivostok (1104 m, 2012), Sutong over the Yangtze River (1088 m, 2011), Stonecutter at Hong Kong (1018 m, 2009), and Tatara in Japan (890 m, 1999).

Figure 1.22 Typical anchor block with splay saddle and bent strut.

Figure 1.23 Shared anchorage at the San Francisco–Oakland West Bay Crossing.

The importance of the cable bent was aptly demonstrated by the failure at Pont des Invalides in 1826. The splay saddle and strut are exposed in Figure 1.22; however, in many cases they are incorporated in the anchorage monolith. The cable bent at the Great Belt is particularly striking because of its clearly defined function and high-profile setting.

Beyond the strand shoes, eyebar chains or rods fan out into the anchorage monolith or into rock along straight lines. At the Brooklyn Bridge (Figure 1.2), the cables enter the anchorage along a horizontal tangent and the cable bent is entirely avoided. The eyebar chains are embedded in the anchor monolith along a 90° arc, reinforced with bearing blocks at every 10°. Gimsing (1997) points out that "after the introduction of prestressing…it has proved advantageous to use post-tensioned bars or cables to transmit the strand forces to the concrete of the anchor block."

Consecutive suspension bridges, such as the two at the San Francisco–Oakland West Bay Crossing (Figures 1.6 and 1.23), the Kita (264/990/264 m) and Minami (254/1048/264 m) Bisan-Seto Bridges, share an anchorage. The three Kurushima Bridges (Figure 1.6) share two anchorages.

1.8 DECK JOINTS AND BEARINGS

The extreme length and flexibility of suspension bridge decks place extraordinary service requirements on their expansion joints and bearings. Decks on most European and Japanese bridges are continuous between the anchorages and suspended at the towers. On American bridges, they are discontinuous at the towers. The latter option requires two expansion

Figure 1.24 Modular and sliding plate joints.

joints and the appropriate sets of bearings per tower, but reduces the movement at the anchorages.

The penetration of water and debris below deck is highly damaging to the sensitive details at towers and anchorages. A variety of joints providing a continuous road surface have been designed as alternatives to the traditional finger joints. Most widespread are the modular joints of various size and displacement capacity. Figure 1.24a is an example. The sliding plate joint, shown in Figure 1.24b, is a model of the joints on the Rainbow Bridge in Tokyo.

Elastomeric, pot, and roller bearings have been used in order to accommodate the large displacements and rotations at supports. Joints and bearings must be designed for high-fatigue stress cycles. They require intensive regular maintenance, consisting of cleaning and lubrication. The high sensitivity to humidity, temperature, displacements, and accelerations makes anchorages, joints, and bearings suitable targets for online health monitoring.

1.9 STRUCTURAL ANALYSIS

1.9.1 The Catenary

In their radically different manner, both Steinman and Ammann acknowledged the role of art and luck in their work. Although recognized as a key ingredient in the engineering of suspension bridges, art endures only with the backing of science. Steinman (1949) states that "the scientific design of suspension bridges dates from about 1898." By that standard, Steinman and Watson (1957) conclude that "John Roebling had built a better bridge than he knew."

Art and science began to converge toward a theory of cable-supported structures, as in most of physics, during the Renaissance. Leonardo da Vinci (1452–1519) speculated about the perfect shapes of both the suspended chain and the voussoir arch. By less than rigorous analogy, Galileo (1564–1642) obtained the shape of a hanging chain as a parabolic arc. Robert Hooke (1635–1703) viewed arches and hanging strings as the reciprocal forms corresponding to pure compression and tension, respectively. The definition of the catenary or funicular as an optimal shape minimizing potential energy owes much to Leonard Euler (1707–1783) and his calculus of variations. According to Irvine (1981), "by the late 17th century the Bernoullis (Jacob [1654–1705] and his brother Johann [1667–1748]), Leibnitz [1646–1716] and Huygens [1629–1695], more or less jointly discovered the catenary." Thus, over roughly 20 centuries, the art and science of bridge building refined their process and expanded their product from compression and the voussoir arch to tension and the suspension structure.

1.9.2 Elastic Theory

Von Karman and Biot (1940), Steinman (1949), Timoshenko and Young (1965), Irvine (1981), and Gimsing (1997) base preliminary estimates of the stresses and shapes of suspension bridge cables on the elastic theory. Steinman (1949, pp. 19–20) stated its five fundamental assumptions as follows:

1. The cable is supposed perfectly flexible, freely assuming the form of the equilibrium polygon of the suspended force.
2. The truss is considered a beam, initially straight and horizontal, of constant moment of inertia and tied to the cable throughout its length.
3. The dead load of the truss and cable is assumed uniform per linear unit, so that the initial curve of the cable is a parabola.
4. The form and ordinates of the cable curve are assumed to remain unaltered upon application of loading.
5. The dead load is carried wholly by the cable and causes no stress in the stiffening truss. The truss is stressed only by live load and by changes in temperature.

Timoshenko (1943) and Timoshenko and Young (1965) analyzed an elastic flexible cable under a vertical load w, uniformly distributed along the horizontal chord with span length ℓ. Equilibrium obtains the tensile force H and the parabolic shape described in Equation 1.6:

$$y = w\,x\,(\ell - x)/(2H) \tag{1.6}$$

at $x = \ell/2$.

$$f = y_{max} = w\ell^2/(8H) \tag{1.7}$$

$$H = w\,\ell^2/(8f) \tag{1.8}$$

where x and y are the abscissa and ordinate of a left-hand coordinate system with origin at the left side support.

The length s of the cable is obtained in terms of ℓ and f by integration and binomial expansion, as shown in Equation 1.9:

$$s = \int_0^\ell \left[1 + (dy/dx)^2\right]^{1/2} dx = \int_0^\ell \left[1 + 64(f\,x/\ell^2)^2\right]^{1/2} dx \approx \ell + 8f^2/(3\ell) \tag{1.9}$$

Differentiating the result of Equation 1.9 with respect to f correlates changes in the sag Δf and in the cable length Δs as follows:

$$\Delta s = 16f\,\Delta f/(3\ell) \tag{1.10}$$

From Equation 1.10, Timoshenko and Young (1965) obtain the sag Δf corresponding to a temperature change $\Delta t°$, as shown in Equation 1.11:

$$\Delta f = c_t \Delta t° \, (3\ell^2 + 8f^2)/(16f) \tag{1.11}$$

where c_t is the coefficient of thermal expansion.

Popular literature about the Golden Gate Bridge quotes that for $\Delta t° = 50°C$, $\Delta f \approx 7$ m midspan.

Small elastic elongation Δs and sag Δf caused by a horizontal force H are obtained similarly as follows:

$$\Delta s = H\,[\ell + 16f^2/(3\ell)]/(A_c\,E_c) \tag{1.12}$$

$$\Delta f = H\,(3\ell^2/16 + f^2)/(A_c\,E_c\,f) \tag{1.13}$$

where A_c is the cross section, f is the sag, and E_c is the cable's elastic modulus.

Timoshenko and Young (1965) obtain H and f for a cable with a horizontal chord, subjected to a superimposed vertical load p, uniformly over limited length $2a < \ell$ and to a concentrated load P.

The authors point out that if p and w represent live and dead loads, respectively, p/w is relatively small. For example, $p/w = 1/6$ at the George

Washington Bridge, presumably without the lower deck. H and f are obtained numerically. Gimsing (1997) further investigates the effect of a horizontal restraint on the cable midspan (as shown in Figure 1.5b).

Without exact knowledge of the preceding information, John Finley recommended practical f/ℓ ratios of 1/6 to 1/7. Gimsing (1997) shows that viable f/ℓ ratios range from 0.1 to 0.15, with 0.25 for just the cable. Theoretically, a cable made of material with design stress $\sigma = 720$ MPa and specific weight $\gamma = 0.09$ MN/m^3 can be expected to carry its own weight up to spans of 10,600 m. By that length, however, the initial assumptions become unrealistic.

Optimizing the global dimensions of a suspension bridge includes the type of tower and stiffening, and the length of the side spans. Cost, local expertise, social demands, and the designer's personal preference are definitive. Table 1.3 summarizes the overall geometry of representative bridges enumerated in Table 1.1. The visual effect of different f/ℓ ratios is illustrated in Figures 1.25 and 1.26. The high f/ℓ ratio of the Golden Gate Bridge appears to enhance its dramatic visual impact.

The 0.58 ratio of the side to main span of the Brooklyn Bridge implies that the decks approach the anchorages above the cables. Roebling countered the greater flexibility entailed by larger ratios of side to main spans with diagonal stays. At the Mackinac, where the ratio of the side to main span is 0.48, Steinman provided an 11.6 m deep space truss. Similar general proportions were selected at the Tagus River Bridge, and again at the Akashi Kaikyo Bridge, where the truss is 14 m deep (Figure 1.4). Side spans are characteristically shorter at Ammann's bridges, up to 0.186 at the George Washington and 0.285 at the Verrazano. Until the lower deck was built in 1962, the George Washington functioned without a stiffening truss, but was stabilized by the superior width and weight of the deck. The depth of the space truss is 7.3 m at the Verrazano Bridge and 7.6 m at the Golden Gate truss (Figure 1.26). The box girder of the Tsing Ma Bridge (Figure 1.5) is 7.4 m deep, to accommodate rail traffic.

Table 1.3 Overall Geometry of Representative Bridges

Bridge	f/ℓ	Side/Main Span	Tower Saddle	Designing Engineer
Brooklyn	0.08	0.58	Rollers	J. Roebling
Williamsburg	0.11	0.37[a]	Rollers	L. L. Buck
Manhattan	0.11	0.49	Fixed	R. Modjeski/L. Moisseiff
George Washington	0.10	0.174, 0.186	Rollers/fixed	O. Ammann
Golden Gate	0.16	0.268	Rollers	J. B. Strauss/C. A. Ellis
Mackinac	0.093	0.48	Fixed	D. B. Steinman
Great Belt	0.11	0.332	Fixed	N. Gimsing
Akashi Kaikyo	0.1	0.482	Fixed	S. Kashima/HSBA

[a] Side spans not suspended.

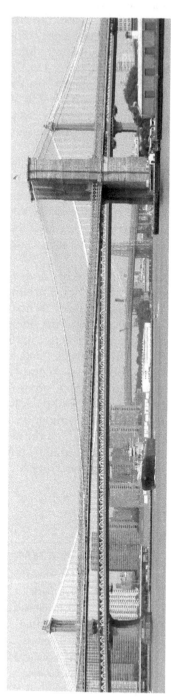

Figure 1.25 Brooklyn, Manhattan, Williamsburg, RFK Triborough (with Hell Gate), Bronx-Whitestone, and Throg's Neck Bridges.   *(continued)*

Figure 1.25 (continued) Brooklyn, Manhattan, Williamsburg, RFK Triborough (with Hell Gate), Bronx-Whitestone, and Throg's Neck Bridges.

Figure 1.26 George Washington and Golden Gate Bridges.

1.9.3 Deck Stiffening

The need for deck stiffening was evident to all successful suspension bridge designers since J. Finley; however, theory and practice were slow to converge on the necessary amount and appropriate means of achieving it. The original constraint on the decks was traffic; however, wind soon proved more formidable. To the consternation of nineteenth-century engineering, the suspension structures were not linearly elastic, not homogeneous, and their behavior was not static. Consequently, attempts to extend suspension spans beyond the empirically confirmed lengths or the service to new usage encountered unforeseen behavior.

By realizing the limitations of the pseudostatic elastic analysis, design managed to remain (in both senses of the term) mostly conservative. Theory eventually grasped both large deflections and dynamic response, unfortunately in that nonconservative order.

1.9.4 Deflection Theory

Steinman (1949, p. 19) cautions that "variations from the 4th assumption of the Elastic Theory may be of sufficient amount to require special consideration as is given by the more exact Deflection Theory," summarized in Appendix D therein. His translation from the German published by J. Melan's theory dating from 1888 was published under the title "Theory of Arches and Suspension Bridges" (1913).

The risks of the gained new knowledge become clear in the following introduction to "Fundamental Equations for Stiffened Suspension Bridges" by Timoshenko and Young (1943): "The deflection of the cable produced by live load is small only in the case of heavy long-span bridges. Otherwise, the deflections may be considerable. In order to reduce these deflections, stiffening trusses are introduced."

The statement could be (and was) misinterpreted to mean that whereas the relatively smaller bridges of Finley and even Roebling needed stiffening, the modern larger spans might not. Timoshenko (1943) ominously comments that "Melan's theory has been widely used in analysis of large-span suspension bridges in this country."

L. S. Moisseiff first applied the theory to the design of the Manhattan Bridge and, later, with F. Lienhard, extended it to lateral forces. Gimsing (1997, p. 15) assessed the development as follows:

> The two-dimensional deflection theory developed by Melan had removed the lower bound for the bending stiffness of the girder in the vertical direction, and now the extension of the deflection theory to cover the three-dimensional behaviour implied that a lower bound for the lateral bending stiffness also disappeared.
>
> In the hands of engineers deprived of their intuitive understanding found in the previous century, and trained to trust blindly the results of their calculations, these analytical achievements could, and should, lead to serious mistakes.

Most serious proved to be the neglect of aerodynamic stability.

1.9.5 Aerodynamic Stability

Whereas approximate static analysis is sufficiently accurate for the preliminary estimates of suspension cable strength, the decks and suspenders add up to a structure with complex dynamic behavior, hard to model theoretically and master practically. As if anticipating the Wheeling collapse in 1854, Roebling (1841) cautioned as follows:

> Railings and longitudinal trusses will not prevent these oscillating motions, but a stiff and well-constructed floor will offer a great resistance. The floors of almost all the English suspension bridges are entirely too light; better specimens in this respect are to be found on the Continent.

Against resonance, apart from his signature diagonal stays, Roebling recommended bringing "the weight of the cables into action by connecting them with the floor at intervals, either by timbers or cast iron pipes, which may include the suspenders."

By the time the Tacoma Bridge failed in 1940, Theodore von Karman (1881–1963) had already presented the airfoil theory for nonuniform motion. After the collapse, Bleich et al. (1950) approached the mathematical modeling of vibrations in suspension bridges as follows:

While the direct stimulus for the formation of the advisory board on the investigation of Suspension Bridges was the failure of the Tacoma Narrows Bridge, it should not be assumed that this was the first occasion wherein dynamic oscillations in suspension bridge structures resulted in damaging stress effects and failure.

The Tacoma failure confirmed that, along with ignorance, misinterpretation and overconfidence are significant vulnerabilities of design. As all major failures, this one was caused by several factors contributing to dynamic instability. Principal among them was flutter.

"Elementary Theory of Wing Flutter" is formulated in Chapter VI. Section 2 of Von Karman and Biot (1940). In Chapter 7, "Flutter Theory," of Bleich et al. (1950), the phenomenon is presented as a form of "self-excitation." Gimsing (1997) defines it as "a harmonic oscillation characterized by a coupling of the vertical and the torsional oscillations occurring when the frequencies of those two basic oscillations coincide." Tacoma became unstable in part because of the shape of its cross section, but also because of its flexibility. Buffeting and vortex shedding may have contributed as well. For members with a dominant mode of oscillation and low damping, resonance is always a threat.

It can be seen from Table 1.1 that practice and analysis eventually converged upon deck stiffening by space trusses (Figure 1.4 and 1.26), and box girders (Figure 1.5). Both systems have been referred to as stiffening girders. Besides the structural performance under the anticipated loads, the choice of one or the other system can be motivated by life cycle maintenance considerations, construction expertise, and aesthetics. Gimsing (1997) considers the introduction of box girders at the Severn Bridge (1966) as "the most important innovation within suspension bridges in the 20th century." They typically imply orthotropic decks, whereas space trusses can be combined with reinforced concrete panels, concrete-filled or open-steel gratings, and so on, as well.

Gimsing (1997) points out examples of aerodynamic stability, achieved by different measures. The Great Belt Bridge, designed mainly to resist wind, has a compact section with all members interacting. The Akashi Kaikyo Bridge, subject to earthquakes as well, has a heavy truss, independent of the deck. The proposed Messina Straights Bridge, which falls somewhere between the two, relies on weight, activates the cable system, and minimizes aerodynamic forces with the shape of the cross section.

The stiffening systems can be discontinuous at the towers, as in most American bridges, or continuous between the anchorages, as in European ones. On many of the latter, such as the Little Belt, the Great Belt (Figure 1.5), and Pont de Tancarville (Figure 1.11), the suspension cables are connected to the stiffening girder midspan with a central clamp. According to Gimsing (1998), a central clamp inhibits the asymmetric torsion mode, transfers axial loads from the girder to the cable, and protects the short suspenders from fatigue.

1.10 VARIATIONS

The suspension and the cable-stayed structural schemes often compete as alternatives for spans of length approaching 1000 m. Under exceptional demands, however, the two systems cooperate and borrow features from each other. Demands qualifying as exceptional have included extreme span length, excessive dynamic loads, environmental constraints, and the desire for uniqueness.

1.10.1 Self-Anchored Bridges

Similarly to bowstring arches and prestressed girders, self-anchored suspension bridges are self-equilibrated structures. The Chelsea Bridge shown in Figure 1.20 is such an example, as are the three chain-link bridges over the Allegheny River in Pittsburgh (Figure 1.8c). The disadvantage of self-anchoring is the need for temporary supports during construction, as shown in Figure 1.7. Consequently, cable-stayed bridges have almost entirely pre-empted this option. The East Bay San Francisco–Oakland Bridge (Figure 1.7) is a notable exception where aesthetic considerations governed.

The vehicular Konohana Bridge (120/300/120 m, 1990) in Osaka (Figure 1.27) is supported by a unique self-anchored monocable.

1.10.2 Hybrid Bridges

Throughout the evolution of cable-supported bridges, hybrids combining stays and suspenders have been proposed whenever the perceived limitations of either system have been exceeded. An early example is the Albert Bridge over the Thames in London, which opened to traffic in 1873 as a cable-stayed structure of the Ordish–Lefeuvre type, was strengthened

Figure 1.27 The self-anchored monocable Konohana Bridge, Osaka. The Albert and Tower Bridges, London.

by a suspension system in 1887, and became simply supported in 1972 (Figure 1.27). The use of trusses as suspension systems before cables had demonstrated their superiority also produced hybrids, notably, the Tower Bridge in London (Figure 1.27). G. Lindenthal's proposal for such a hybrid across the Hudson River was rejected in favor of O. Ammann's George Washington Bridge.

Roebling's Brooklyn Bridge remains the iconic representative of the hybrid suspension–stay system worldwide. In tribute to J. Roebling, J. Schleich, and R. Walther proposed hybrid twin bridges (Walther and Amsler, 1994) for a replacement of the Williamsburg Bridge. A hybrid bridge was considered for the East Bridge crossing in Denmark before the Great Belt suspension bridge was selected. The Third Bosphorous Bridge is a suspension–stay hybrid designed for mixed vehicular and rail traffic.

Historically, stays have either been concurrent with the suspenders throughout the spans (as in Roebling's designs) or acted as sole supports of portions of the spans near the towers. In 1938, Dischinger proposed a combined cable and stay system in which stays fan out in the proximity of the towers toward the deck, and suspenders support only the center of the span. Gimsing (1997) comments that "strangely enough, although Dischinger adopted the idea of combining the suspension system and the cable stayed system, he did not appreciate the original solution of Roebling with the much more continuous lay-out."

Gimsing (1997) attributes to D. Steinman a proposal for a suspension bridge at the Messina Straits with "negative" stays radiating from the towers at deck level toward the cables in the central span. The idea can be extended to a cable net system, combining suspenders, stays, and secondary or trajectory cables. Steinman designed the Tagus River Bridge in Lisbon (632/1013/632 m, 1966) for vehicular traffic on its original upper deck, but anticipating the addition of a lower deck for two train tracks. In contrast with O. Ammann's George Washington Bridge, however, the Tagus would have been reinforced by diagonal stays, thus making it a hybrid. Instead, in 1991, the bridge towers were heightened and two new main cables were added.

1.10.3 Multispan Systems

Four- and five-span alternatives were considered for the San Francisco–Oakland West Bay Bridge, before the two consecutive suspension bridges (Figure 1.23) were selected. In the absence of a shared anchorage, a suspension bridge with more than three spans (and two towers) is multispan. Such bridges were much more common in the nineteenth century than during the twentieth. The most important example of this type was the 5×109 m span bridge at Cubzac (1835–1869) over the Dordone in France (Marrey, 1990). Its ultimate closure was reportedly caused by scour at one of the foundations.

Figure 1.28 (a) Sunniberg Bridge, Switzerland. (b) Viaduct de Millau, France.

The potential instability of intermediate towers in multispan bridges was remarked on by Navier (Kranakis, 1997). Gimsing (1997) demonstrated that a four-span configuration would develop inadmissible deflections unless stiffened by supplementary means, such as horizontal cables between tower tops, or the more common diagonal stays to the deck level of opposite towers, as at Cubzac and, more recently, the four-span cable-stayed Ting Kau Bridge in Hong Kong (127/448/475/127 m, 1998).

Nets of inclined suspenders supported the five-span San Marcos "cable truss" bridge in El Salvador (76/159/204/159/76 m). The spectacular multispan cable-supported bridges accomplished in recent years have been extrados, such as the Sunniberg in Switzerland (Figure 1.28a), or cable stayed, such as the Viaduct de Millau in France (Figure 1.28b) and Rion-Antirion in Greece. Significantly, the construction of the Viaduct de Millau also required temporary intermediate supports.

Technological advances have revived even steel plates as continuous-tension elements. Passerelle Simone de Beauvoir (304 m, 2006) over the Seine (Figure 1.29) is a lenticular pedestrian bridge supported by two continuous steel plates (1000/150 mm) acting predominantly in tension, as cables might have done.

The Taizhou Bridge over the Yangtze River (2012) has two suspended spans of 1080 m (3540 ft) length. The central steel tower is 192 m (630 ft), and the two concrete side towers are 178 m (584 ft) high. Two 390 m (1279 ft) long side spans are supported on multiple piers. The suspension cables are two. The deck box girder carries six lanes of vehicular traffic.

Figure 1.29 Passerelle Simone de Beauvoir, Paris. Insets: Anchorages of the Passerelle and Pont des Invalides.

1.11 LESSONS

Whereas art masterpieces cannot be reproduced, science advances through failures toward understanding. The design and construction of suspension bridges owe much to the lessons of instructive failures.

1.11.1 Failures

Sibly and Walker (1977) and Petroski (1993) discern a cyclic trend in bridge failures at the Dee Bridge in England (1847), the Tay Bridge in Scotland (1879), the Quebec Bridge (1907), and the Tacoma Narrows Bridge (1940). The cycles begin with cautiously successful (partly empirical) designs and culminate by overextending the practice beyond the validity of the model. Petroski (1994) presents the failure of the Dee Bridge at Chester as an example of the "success syndrome," essentially an error of complacency. Structural failures of this type result from applying known routines beyond their valid range. Suspension bridges are particularly vulnerable because their span and function are always at the limit of the technically possible range. After the failure of the Tacoma Narrows Bridge, Ammann (1953) wrote: "[It] has given us invaluable information.... It has shown [that] every new structure [that] projects into new fields of magnitude involves new problems for the solution of which neither theory nor practical experience furnish an adequate guide. It is then that we must rely largely on judgment and if, as a result, errors, or failures occur, we must accept them as a price for human progress."

An early explorer of suspension systems, Leonardo da Vinci (1935), wrote in his *Notebooks*: "Experience is not at fault, it is only our judgement that is in error."

Kranakis (1997) demonstrates how the interpretation of failure causes has evolved concurrently with and sometimes independently from the empirical elimination of incalculable risk.

The following examples continue to advance the art and science of bridge engineering by demanding continuing reinterpretation with every new generation of practitioners.

1.11.1.1 *Pont des Invalides (1826)*

Navier attributed the failure of Pont des Invalides over the Seine in 1826 to poor material in the anchorages. His drawings suggest a precarious overturning moment as well (Figure 1.29, inset). The massive rigid anchor monoliths of later anchorages (Figure 1.22) eliminate both deficiencies.

Twenty-first-century technology overcame Navier's limitations at the Passerelle Simone de Beauvoir in Paris (Figure 1.29, inset). The geometry of the Passerelle anchorage vaguely resembles the one at Invalides, the better to expose the critical differences. The masonry compression strut is replaced by steel. The tension element approaches the anchorage at an acute, rather than an obtuse, angle. Highly compressed flat arches aboveground and highly pretensioned strands below grade resist the overturning moment. Whereas the nineteenth-century anchorage was entirely passive, the twenty-first-century one can be viewed as, if not active, then at least responsive. Maintaining that response becomes a life cycle responsibility.

1.11.1.2 *Wheeling Bridge (1848–1854)*

Over a period of 86 years after the collapse of this world's longest span, the need for deck stiffening, by Roebling's hybrid system or by trusses, remained empirically obvious. Then the lesson had to be reformulated and relearned analytically.

1.11.1.3 *Tacoma Narrows Bridge (1940)*

The failure of the Tacoma Narrows Bridge stimulated both theory and empiricism. The results obtained by the deflection theory were recognized as potentially nonconservative. Von Karman's analysis of airplane wing aerodynamic stability was applied to bridges. Space trusses, box girders, and inclined suspenders henceforth compete as means of deck stiffening.

Large-scale physical models of entire bridges or deck sections are tested extensively in wind tunnels. For the wind tunnel test of the Tsing Ma

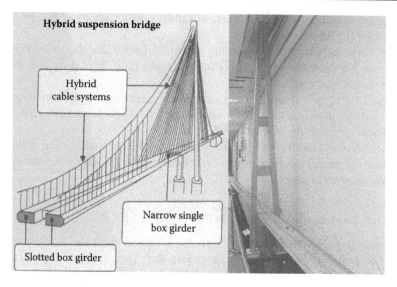

Figure 1.30 Study of hybrid bridge (1100/2800/1100 m), PWRI, Tsukuba City.

Bridge, a 300 m long section was modeled at a scale of 1/90. A 1/100-scale model for the full-length Akashi Kaikyo Bridge was tested in a 41 m wide wind tunnel built for the purpose at Tsukuba City.

Around 1820, the Seguin family tested an 18 m long, 0.5 m wide model of their suspension bridge at Marc Seguin's property near Annonay (Marrey, 1990). Models of hybrid bridges with a total length of 5000 m are tested as of this writing (Figure 1.30).

Weigh-in-motion systems have been installed or are considered for many long-span bridges for accurate assessment of the impact of overweight vehicles.

1.11.1.4 Point Pleasant Bridge (1928–1967)

The Silver Bridge over the Ohio River at Point Pleasant failed in December 1967, due to the brittle fracture of one critically nonredundant eyebar. An uninspectable material defect had initiated the fracture. In direct response, the U.S. Congress launched the National Bridge Inventory (NBI), and thereby modern bridge management.

Recognizing the superiority of high-strength steel wire cables over eyebar chains implies several important lessons:

- Cost is shown to be an insufficient criterion in the selection of essential structures with long spans in space and perpetual life cycles in time.
- Redundancy is understood to imply not only static indeterminacy, but also a viable path of load redistribution. Internal redundancy and ductility become increasingly associated. Fracture-critical structural elements are targeted for elimination.

- John Roebling's singular blend of practical judgment and scientific knowledge reemerges as the prototype for a successful integration of the process and product of the suspension bridge. His designs prefigure the contemporary notions of robustness and resilience. His hybrid suspension–stay system remains relevant every time spans exceed customary lengths and stiffening becomes critical. Most major malfunctions of suspension bridges can be traced to ignoring one or more of Roebling's lessons.

1.11.2 Partial Failures

Another definition of structural failure is nonperformance. Partial performance amounts to partial failure. With the opportunity to examine nonperformance over time come the responsibilities to arrest its progress and eliminate its cause.

Many important improvements and innovations in the design, construction, and management of suspension bridges have been inspired by partial failures. In such cases, it is of great benefit to determine what has prevented the total failure. On the conceptual level, the Brooklyn Bridge and Roebling once again supply the example. Roebling's response to the criticism of his heterogeneous means of load distribution was that if any of his systems fails, the bridge "may sag but shall not fail." Billington (1983) speculates that, lacking the analytic tools, Roebling obtained a working stiffening scheme formalistically. The rigorously analytic Navier had rejected the combination of suspenders and stays (Kranakis, 1997).

Roebling's reasoning in favor of redundancy and robustness was confirmed in 1981 when a diagonal stay with an estimated weight exceeding a ton fell and killed a pedestrian on the Brooklyn Bridge promenade. All stays and suspenders had corroded nearly to failure. Despite the advanced state of deterioration, however, the entire system was replaced under traffic because of its redundancy, as design had anticipated.

Throughout the twentieth century, parallel wire cables have demonstrated a considerable resilience, even when their protection from corrosion has been neither redundant nor robust. The cable wires of the Williamsburg Bridge had not been galvanized on the strength of the argument that if they were not protected, zinc would not save them, and if they were, zinc would be redundant. Besides savings from the deferred galvanization, the benefit was a lighter cable. In 1988, strands of the main cables were found corroded to failure. One strand was broken in the Manhattan Anchorage. Conservative design saved the bridge. The four cables themselves had been designed with a robust strength reserve estimated at 4.2.

Whereas a crack initiation in a fracture-critical element is hard to spot before it causes a global failure, wire breaks in a properly designed and inspected cable-supported structure, as in the cases shown in Figure 1.31, can be treated as localized failures with a reversible effect. Taking into account

Figure 1.31 Broken wires in a main cable and in a suspender rope.

the observed corrosion (Figure 1.9), as well as internal stress redistribution through friction, the strength reserve of the Williamsburg Bridge cables was found to have dropped to a still acceptable factor of approximately 3. Future maintenance would have no further tolerance for partial failures.

Corrosion of anchorage eyebars necessitated the reanchoring of cable strands at Manhattan, Williamsburg, RFK Triborough, George Washington, and Bronx-Whitestone Bridges. A partial reanchoring of a strand is illustrated in Figure 1.32. The corrosion of the original anchor is attributed to leakage of the finger joints above the anchorage and to lack of dehumidification

Figure 1.32 Reanchoring of a strand.

within. Thus, once again, the product and the process are both at fault for relying excessively on each other.

Certain missteps in the original concept or execution emerge only after years of service. For example, the straps compacting the cable strands, as shown in Figure 1.21a, can have a corrosive effect if made of mechanically intrusive or chemically active material. The latter was true of the original bronze encasement in Figure 1.21c.

Helical strand cables are particularly vulnerable to corrosion concealed within the strands. Since a partial failure manifests itself by a broken strand, rather than individual wires, the strands themselves can be viewed as fracture-critical. The Waldo-Hancock Bridge was demolished in 2013. The new helical strands at Pont de Tancarville (1998) and Pont d'Aquitaine (2002) are galvanized but not locked-coil ones.

As all significant structures, suspension bridges fail because of more than one cause. Failures in the product and the process are investigated by forensics and management, respectively. If the two types of investigations do not converge, their findings also fail to achieve full impact. No structure demonstrates better than a suspension bridge the critical importance of continuity within, as well as between, the process and the product.

The Kukar suspension bridge (100/270/100 m) in Indonesia collapsed on November 26, 2011. At least one analysis has argued that the failure had "plural sources but a single cause." The following facts are reported. At least one anchorage block, sitting on vertical piles, slipped. The tower tilted and the main span sagged. To correct the sag, one of the two trusses was jacked up by midspan suspender 13 (possibly by 10 cm). The cast-iron clamp of the suspender to the main cable fractured, followed by a fracture of the symmetric one on the opposite cable. The remaining clamps, spaced at 10 m, failed successively. Prior to the failure, the suspenders were experiencing increasing vibrations. The two main cables consisted of 19 helical strands. Slippage of the cable clamps had been reported. The bridge had been in service for 10 years.

The effort to identify and prevent causes for partial and total failures from developing in the process and the product has engendered the notions of vulnerabilities and potential hazards. Under the term *risk*, Hovhanessian and Laurent (2006) seek vulnerabilities in critical elements of the product (i.e., the suspension bridge) and in the key stages of the process, comprising design, construction, aging, operation, hazards (i.e., extreme events), and maintenance. Previously vague considerations known to avert potential failures and improve structural performance have become explicit and specific, as, for example, the following.

1.11.2.1 Redundancy/Robustness

The inability to redistribute functional demands makes the difference between the partial and total failure. Thus, robustness can be perceived as a redundancy extended beyond the alternate load paths to encompass unforeseen functional demands. Bridge elements, including suspenders and stays, as well as deck panels, have to be designed for maintenance and eventual replacement without service interruption, as well as for the partial failures of these operations.

1.11.2.2 Inspectability/Maintainability

The inaccessible is unmanageable. Since the main suspension cables cannot be accessible throughout their cross section, particular attention must be paid to their maintenance. The experience with the Williamsburg Bridge inspired two types of countermeasures.

Under project FHWA-HRT-14-024, Columbia University monitored the humidity, temperature, corrosion, and other parameters in a 10,000-wire cable in laboratory conditions and on the Manhattan Bridge, in order to provide online information about the ambient level of corrosiveness. The Honshu-Shikoku Bridge Authority (HSBA) developed the method of dry-air injection (Figure 1.33) under a wrapping of z-shaped high-strength wire (Figure 1.9e). By reducing the cable humidity to 40%, the system precludes any corrosion. The cable dehumidification method is gaining application worldwide. Most anchorages are now considered maintenance-intensive areas and their dehumidification is routine.

1.11.2.3 First/Life Cycle Cost

After the financial constraints of the 1930s precluded a proposed second level at the Triborough Bridge, Robert Moses reportedly commented that in 40 years, New Yorkers would build a bigger bridge anyway. Eighty years later, the indispensable RFK Triborough Bridge was rehabilitated extensively under daily traffic of up to 200,000 vehicles.

Figure 1.33 Cable dehumidification at the Kurushima Bridge.

Figure 1.34 The Bronx-Whitestone Bridge, 1990.

At the time of their construction, record-breaking suspension bridges, similarly to tall towers, generate an excitement more appropriate for competitive sports. Once in service, however, they become fixtures of the regional topography and shortcuts in the social brain. Cities cannot exist and countries are not neighbors without them. After more than a century, the required service changes in nature and increases in volume, rendering a closure unacceptable. A million users cross the three East River suspension bridges (Figures 1.2 and 1.25) daily. This comes at a price. Over the last 20 years, the same three bridges have absorbed over US$3 billion in rehabilitation projects. Maintenance costs are supplemental.

The Bronx-Whitestone Bridge (1939), originally similar to the Tacoma Narrows Bridge, was equipped with diagonal stays, stiffening trusses (1946), and a tuned mass damper (1985) (Figure 1.34). Ultimately, all added features were to be scrapped in favor of a modified girder profile and stiffening. Modifications to the roadways, the cables, and the anchorages are under review (Lorentzson et al., 2006). The primary design constraint is clearly not the bridge, but the essential service it has provided for more than 70 years.

The lower deck of the George Washington Bridge was installed in 1962, to accommodate the growing demand for vehicular traffic. The average daily traffic (ADT) reported in 2011 was 276,150 vehicles.

Considerations about the full replacement of the Williamsburg Bridge in 1988 showed that despite the billion-dollar price tag, rehabilitation was preferable if user costs were taken into account.

If suspension bridges are designed for a perpetual life cycle, periodic upgrading must be part of it. Partial rehabilitation is a certainty.

1.11.2.4 Function/Form

While artists are balancing form and function, scientists are reconciling determinism and uncertainty. The form of long-span suspension bridges is determined by natural constraints. Function remains uncertain throughout the life cycle. Unforeseen vulnerabilities, as well as benefits, may surface early on or long after a span has made its mark in the rankings for unsupported length.

In the early twenty-first century, as in the nineteenth, random winds and earthquakes, as well as predictable of pedestrians, still excite suspension bridges beyond expectations. In 2000, the torsional stiffness of the three-span Millennium Bridge in London (Figure 1.35), which combines suspension and extrados features, had to be corrected in order to accommodate self-synchronizing pedestrians.

In a statement similar to Roebling's (1841) argument about the importance of science, Eiffel famously argued that his tower would be beautiful because the equations that describe it are correct. At the end of his spectacular career, Ammann (1879–1966) identified luck as his best asset (Talese, 1964). The creativity of these masters, however, is revealed not in their words, but in the unity between their product and process. There is no separation or conflict between the beauty and utility of their structures. The success of suspension bridges at the extreme edge of the possible depends on their organic integration of process and product, form and function, design and performance. Hence, a review of their evolution may contribute to cultivating a taste for it.

Figure 1.35 Millennium Bridge, London.

REFERENCES

Ammann, O. (1953). *Present Status of Designs of Suspension Bridges with Respect to Dynamic Wind Action*, 231–255. Boston Society of Civil Engineers, Boston.

Billington, D. P. (1983). *The Tower and the Bridge*. Basic Books, New York.

Bleich, F., C. B. McCullough, R. Rosecrans, and G. S. Vincent. (1950). *The Mathematical Theory of Vibration in Suspension Bridges*. Bureau of Public Roads, Department of Commerce, Washington, DC.

Durkee, J. (1966). Advancements in suspension bridge cable construction. In *Proceedings of the International Symposium on Suspension Bridges*, 425–449. Lisbon.

Gimsing, N. J. (1997). *Cable Supported Bridges*. 2nd ed. John Wiley & Sons, New York.

Gimsing, N. J., ed. (1998). *The East Bridge*. A/S Storebæltsforbindelsen, Copenhagen.

Gjelsvik, A. (1991). The development length of an individual wire in a suspension bridge cable. *Journal of Structural Engineering*, 117(4), 1189–1201.

Gourmelon, J.-P., and Brignon, eds. (1989). *Les ponts suspendus en France*. Le Laboratoire Central des Ponts et Chaussées (LCPC) et Le Service d'Études Techniques des Routes et Autoroutes (SETRA), Paris.

Honshu-Shikoku Bridge Authority. (1998). *The Akashi-Kaikyo Bridge: Design and Construction of the World's Longest Bridge*. Honshu-Shikoku Bridge Authority, Kobe, Japan.

Hovhanessian, G., and E. Laurent. (2006). Instrumentation and monitoring of critical structural elements unique to suspension bridges. In *Advances in Cable-Supported Bridges*, 111–120. Taylor & Francis Group, London.

Irvine, M. (1981). *Cable Structures*. MIT Press, Cambridge, MA.

Kranakis, E. (1997). *Constructing a Bridge*. MIT Press, Cambridge, MA.

Leonardo da Vinci. (1935). *Notebooks*, ed. E. McCurdy. Empire State Book Co., New York.

Lorentzson, J., P. Nietzchmann, G. Fanjang, and C. Gagnon. (2006). Planning and engineering for the future: Capacity increase and cable replacement at the Bronx-Whitestone Bridge. In *Advances in Cable-Supported Bridges*, 145–162. Taylor & Francis Group, London.

Marrey, B. (1990). *Les ponts modernes, 18e–19e siecles*. Picard Éditeur, Paris.

Mayrbaurl, R. (2006). Wire test results for three suspension bridge cables. In *Advances in Cable-Supported Bridges*, 127–142. Taylor & Francis Group, London.

Mayrbaurl, R., and S. Camo. (2004). Guidelines for the inspection and strength evaluation of suspension bridge parallel-wire cables. Report 534. National Cooperative Highway Research Program, Transportation Research Board, Washington, DC.

Nishino, F., T. Endo, and M. Kitagawa. (1994). Akashi Kaikyo Bridge under construction. Complementary report presented at the 7th Structures Congress, Atlanta, GA, April 24–28.

Petroski, H. (1993). Predicting disaster. *American Scientist*, 81, 110–113.

Petroski, H. (1994). Success syndrome: The collapse of the Dee Bridge. *Civil Engineering*, 64(4), 52–55.

Reier, S. (1977). *The Bridges of New York*. Quadrant Press, New York.

Roebling, J. (1841). Some remarks on the suspension bridges, and the comparative merits of cable and chain bridges. *American Railroad Journal and Mechanics' Magazine*, XII(379), 193–196.

Scott, R. (2001). *In the Wake of Tacoma*. ASCE Press, Reston, VA.

Sibly, P., and A. S. Walker. (1977). Structural accidents and their causes. In *Proceedings of the Institution of Civil Engineers*, London, vol. 62, part 1, pp. 191–208.

Steinman, D. B. (1949). *A Practical Treatise on Suspension Bridges*. 2nd ed. John Wiley & Sons, New York.

Steinman, D. B., and S. R. Watson. (1957). *Bridges and Their Builders*. Dover Publications, New York.

Strauss, J. S. (1938). The Golden Gate Bridge. Report of the chief engineer to the board of directors of the Golden Gate Bridge and Highway District, California. 50th anniv. ed., 1987.

Talese, G. (1964). *The Bridge*. Harper and Row Publishers, New York.

Timoshenko, S. (1943). Theory of suspension bridges. *Journal of the Franklin Institute*, 235(4), 327–440.

Timoshenko, S., and H. D. Young. (1965). *Theory of Structures*. 2nd ed. McGraw-Hill, New York.

Von Karman, T., and M. A. Biot. (1940). *Mathematical Methods in Engineering*. McGraw-Hill, New York.

Waddell, J. A. L. (1916). *Bridge Engineering*. John Wiley & Sons, New York.

Walther, R., and D. Amsler. (1994). Hybrid suspension systems for the very long bridges: Aerodynamic analysis and cost estimates. In *Cable-Stayed and Suspension Bridges*, 529–536. Association Française Pour la Construction, Deauville, France.

Chapter 2

Design and Construction of Suspension Bridges

Katsuya Ogihara

CONTENTS

2.1 INTRODUCTION

A suspension bridge is a type of bridge in which a floor structure is hung from suspension cables. This type of bridge consists of main cables, stiffening girders, suspender ropes that hang stiffening girders from main cables, main towers that support main cables, and anchorages that anchor main cables.

A suspension bridge is a very flexible structure compared with other types of bridges, and prone to be vibrated by the vehicles running on it and dynamic forces such as wind or an earthquake.

When live loads (vehicular load or train load) act on the stiffening girder, the force is transmitted from the stiffening girder via the suspender ropes to the main cables; the main cables deform and reach a state of equilibrium. Tensile force in the main cable is the dominant factor in the deformation of the main cable. While the stiffness of the ordinary girder bridges is controlled by the area or moment of inertia of the cross section of the girder, that of the suspension bridge is mainly controlled by the dead load.

When a suspension bridge has larger cable sag, tensile force in the main cables is small, and therefore the stiffness of the bridge is small. The large cable sag requires high main towers, and the low stiffness is disadvantageous in terms of wind-resistance design. On the other hand, tensile force in the main cable is large and stiffness of the bridge is large when cable sag is small. In this case, the bridge requires large main cables. Therefore, cable sag of a suspension bridge is decided considering the stiffness and the cost of the bridges.

2.2 STRUCTURAL SYSTEM

2.2.1 Structural Components

A general suspension bridge system is shown in Figure 2.1. The following list explains functions of each member in the system.

1. Stiffening girders: Longitudinal structures that support and distribute live loads and secure the aerodynamic stability of the structure.
2. Main cables: Parabolic tensile structure that supports stiffening girders by suspender ropes and transfers the force to main towers.
3. Main towers: Vertical structures that support vertical load from the main cables and transmit the force to the foundations.
4. Anchorages: Structures that resist tensile force of the main cables.

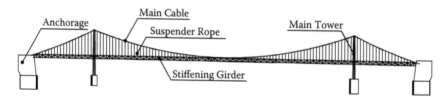

Figure 2.1 Suspension bridge system.

2.2.2 Types of Suspension Bridges

Types of suspension bridges can be classified by the number of spans. As shown in Figure 2.2, there are single-span, three-span, and multiple-span suspension bridges. Three-span suspension bridges are the most common and adopted worldwide. Single-span suspension bridges can be considered one version of three-span suspension bridges since they are supported by two main towers. Only their center spans are hung from the main cables. This type is occasionally selected because of topographic restrictions. Multiple-span suspension bridges are supported by more than two main towers. When one main span is loaded with live load, the main tower is considerably deformed, and hence the deformation of the stiffening girders becomes significant. Also, since the natural frequency of a bridge becomes small, it is unfavorable in terms of wind-resistance design. Because of these downsides, this type is rarely chosen. The newly built Oakland–San Francisco Bay Bridge features two main spans supported by one main tower (Nader et al. 2013).

Another way of classification is whether stiffening girders are hinged or continuous at the main towers. For suspension bridges that support vehicular traffic, hinged stiffening girders are usually adopted. For suspension bridges that support railways, continuous stiffening girders are selected in order to secure the smooth operation of trains.

For ordinary suspension bridges, main cables are anchored at anchorages that are independent from the superstructures. A self-anchored suspension bridge is a type of suspension bridge in which the main cables are anchored to the stiffening girders at its ends. In this type, axial force is introduced to the stiffening girder.

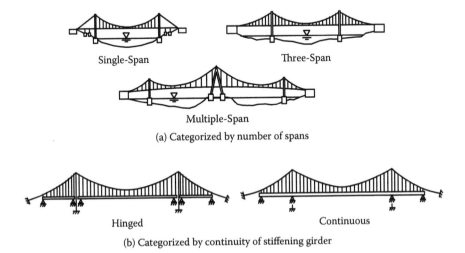

(a) Categorized by number of spans

(b) Categorized by continuity of stiffening girder

Figure 2.2 Types of suspension bridges.

2.2.3 Main Towers

It is a common practice to design main towers not to resist the longitudinal displacement of the main cables. Therefore, for small- or middle-sized suspension bridges, main cables are supported by the longitudinally movable saddles at the tower tops, or rocker-type towers are adopted. For long-span suspension bridges, main towers are usually designed as flexible towers. Although the base of a main tower is rigidly fixed at the bottom and the cable movement is not allowed at the tower top, the main tower does not resist the cable movement because of its flexibility.

In the transverse direction, main towers are usually composed of two shafts to accommodate stiffening girders in between. And to resist transverse forces such as wind, two shafts are usually connected to enhance rigidity. Towers with diagonal bracing have higher rigidity than portal towers.

Types of main towers are shown in Figure 2.3.

2.2.4 Main Cables

In the early suspension bridges, main cables were composed of chains or eyebars to achieve flexibility, and the material was mainly wrought iron. For modern suspension bridges, steel is used to achieve high strength, and hence reduce the total weight of the superstructure. Steel main cables are categorized largely into wire rope cables and parallel wire cables, depending

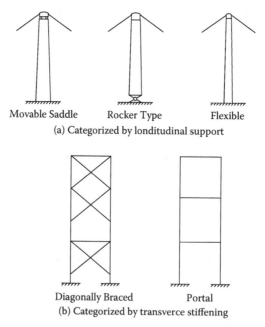

Movable Saddle   Rocker Type   Flexible

(a) Categorized by londitudinal support

Diagonally Braced   Portal

(b) Categorized by transverce stiffening

Figure 2.3 Types of main towers.

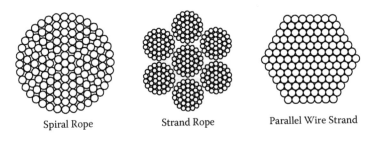

Spiral Rope   Strand Rope   Parallel Wire Strand

Figure 2.4 Types of components of main cables.

mainly on the size of the bridge and the time when they were built. Wire rope cables are composed of twisted ropes arranged in parallel, such as stranded ropes or spiral ropes, and used for relatively smaller suspension bridges. Parallel wire cables are composed of parallel wires and categorized by the construction methods, such as the air-spinning (AS) method and prefabricated parallel wire strand method. The parallel wire cables were used for the Brooklyn Bridge for the first time and are used for most modern suspension bridges.

Figure 2.4 shows types of main cable components.

2.2.5 Suspended Structures

Suspended structures are all the structural components that are hung from main cables and consist of suspender ropes, stiffening girders, and decks. Since the stiffness of the suspension bridge is mainly controlled by the cable tension, stiffening girders can be considered secondary structures that give additional stiffness to the bridge. Therefore, it is possible to keep the cross section of the members small, which is required to secure the stiffness of the bridge and support live loads.

As the design theory of suspension bridges advanced, the cross section of the stiffening girders diminished. Especially after the finding of the deflection theory, plate girders were used in order to reduce the weight. However, the collapse of the Tacoma Narrows Bridge let bridge engineers realize that a certain amount of stiffness is required to ensure stability against strong wind. After the Tacoma collapse, truss girders were mainly used for the stiffening girders that give appropriate rigidity and aerodynamic characteristics. The advancement of the wind-resistant design gave more freedom to the bridge engineers. The box-shaped girder was first introduced for the Severn Bridge after thorough investigation, and it is used for modern suspension bridges worldwide.

Figure 2.5 shows types of stiffening girders.

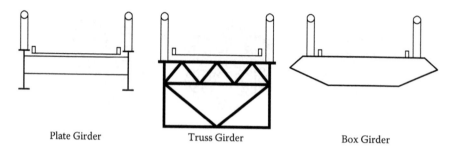

Plate Girder   Truss Girder   Box Girder

Figure 2.5 Types of stiffening girders.

2.2.6 Anchorages

As mentioned in Section 2.2.2, main cables are usually anchored at the anchorages that are independent of other structural components, except for self-anchored suspension bridges. Since it is difficult to anchor thick main cables at a single position, the main cables are spread into smaller bundles, usually called strands, to be anchored separately. Anchorages are categorized by the resisting mechanism against large cable tension (Figure 2.6). Gravity-type anchorages are the most common, and they resist cable force by their weight. In tunnel-type anchorages, strands are anchored in the anchor block embedded into sound bedrock. The friction between the anchor block and the surrounding bedrock resists the cable force. In case the bedrock is in very good condition, the strands are directly anchored in these bedrocks. This type of anchorage can reduce the amount of excavation and therefore is economical.

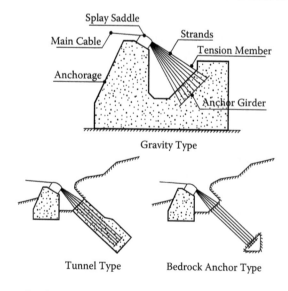

Figure 2.6 Types of anchorages.

2.3 DESIGN

2.3.1 General

The first theoretical design method of the suspension bridge was issued in 1823 by Navier as the elastic theory. The elastic theory was dominant for the design of suspension bridges in the first half of the nineteenth century. However, since the suspension bridge is a highly flexible structure, its nonlinearity is inherently very high. The elastic theory that considers the stiffness of the whole bridge without deformation by live loads gives safer side approximation and was considered uneconomical. In 1888, Melan issued the deflection theory that considered the deformed state of the suspension bridge by live loads. According to the deflection theory, the stiffness of the stiffening girder can be reduced dramatically because the moment induced by the live loads can be reduced by the deformation of the bridge itself. Until this time, the dynamic effects of wind force were not seriously considered in suspension bridge design. As the study of suspension bridges advanced, wind effect became a critical factor in suspension bridge design. The deflection theory or its versions with appropriate linear approximations were used until the 1960s. Today, with the advancement of computer technology and the theory of numerical structural analysis, the finite element method that can take nonlinearity into account is broadly used.

2.3.2 Classic Analytical Methods

2.3.2.1 Elastic Theory

The elastic theory is based on the infinitesimal deformation theory. In this theory, the following are assumed:

1. Main cables are fully elastic and do not resist bending.
2. The stiffening girder is horizontal, and the moment of inertia is constant. It is considered a beam that is supported by main cables along its entire length.
3. Dead loads of the stiffening girder and main cables are uniformly distributed. Therefore, the cable shape is parabolic.
4. Main cables do not deform even under loading.
5. All dead loads are supported by the main cables. Stress is induced in the stiffening girder only by the live load or thermal load.

Bending moment M of the stiffening girder by live load is given by the following equation. The contribution of the main cable appears in the second term.

$$M = M_0 - H_p y(x) \tag{2.1}$$

where M_0 is the bending moment induced in a simply supported beam with the same span length as the stiffening girder by live load, H_p is the horizontal tensile force of the main cable by live load, and $y(x)$ is the vertical coordinate of the main cable.

The elastic theory can be applied to the suspension bridges that have relatively short spans and stiffening girders with relatively high stiffness.

2.3.2.2 Deflection Theory

The deflection theory solves a differential equation that is derived from the equilibrium of the main cable after the deformation. All the assumptions used for the elastic theory are used except number 4. Assumption 4 can be rewritten as follows:

4'. Main cables deform in the vertical plane under loading.

Bending moment M of the stiffening girder by live load is given by the following equation. Contribution of the deformation appears in the third term.

$$M = M_b - H_p y(x) - (H_W(x) + H_p)\eta(x) \tag{2.2}$$

where η is the deformation of the main cable by live load, and H_W is the horizontal tensile force of the main cable by dead load.

The concept of the deflection theory is shown in Figure 2.7. In the deflection theory, one more equation is required since the elongation of the main cables is considered. A condition that the distance of the main cable between the anchorages is constant is supplied. By this condition, thermal variation can be treated as well as live load.

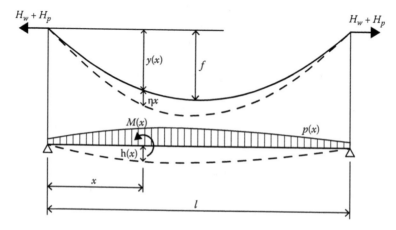

Figure 2.7 Concept of deflection theory.

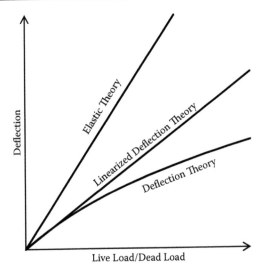

Figure 2.8 Elastic theory, deflection theory, and linearized deflection theory.

Since the deflection theory is to solve nonlinear equations, it requires iteration, and therefore was tedious in the era when computer technology was immature. The linearized deflection theory is a simplified form of the deflection theory and gives reasonable results if the live load/dead load ratio is very small, such as in long-span suspension bridges (Figure 2.8). The influence line method can be used since it is a linear equation.

2.3.2.3 Analytical Method for Lateral Loading

For the suspension bridge under lateral load, such as wind, Moisseiff and Lienhard (1933) established an analytical method. In this method, wind load is considered to be distributed to the main cable and stiffening girders. The elastic equations for the deformation of main cables and that of stiffening girders are solved simultaneously with compatibility conditions of main cables.

2.3.3 Design Procedure

The general design procedure for suspension bridges is shown in Figure 2.9. In the beginning, basic conditions that need to be known for the design, such as natural or environmental conditions, are investigated. Using these basic conditions, an initial design that includes bridge type, configuration, and property of members is assumed, and analytical models are constructed. In-plane analysis, out-of-plane analysis, and seismic analysis are conducted for live load, wind load, and effects of earthquakes, respectively. For these analyses, the classical theories outlined in the previous section were used

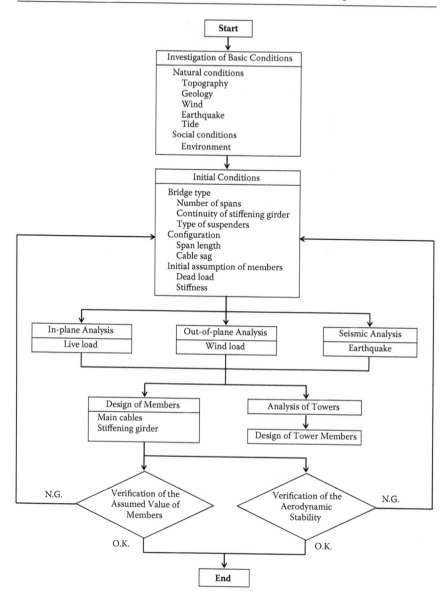

Figure 2.9 Design procedure of suspension bridge.

for early suspension bridges. Today, thanks to the advancement of computer technology, these analyses are generally conducted by the finite element method. Main towers are analyzed separately from the cable–girder system since they are considered not to restrict the movement of the main cables. The displacement of cables and reaction forces of cables and girders obtained from the previous cable–girder analysis are used as inputs for the design of main towers. For long-span suspension bridges, stability against strong wind is the most critical factor. Therefore, if the initial design does not satisfy the required aerodynamic stability, the design must be reconsidered.

2.3.4 Wind-Resistant Design

As mentioned in the previous section, wind-resistant design is the most important factor in the design of the long-span suspension bridge. To enhance the aerodynamic stability, there are two approaches:

1. Select an aerodynamically stable cross section of the member.
2. Improve dynamic characteristic of the structure.

The following is the procedure of the wind-resistant design conducted for the Honshu-Shikoku bridges (Honshu-Shikoku Bridge Authority 1990):

1. Static analysis with assumed aerodynamic coefficients. For the lateral static analysis by the wind load, assumed aerodynamic coefficients are used to calculate wind force acting on the members. Usually these coefficients are derived from the existing data.
2. Selection of aerodynamically stable cross sections. By the wind tunnel tests with sectional or partial models, aerodynamically stable cross sections of the members are selected. For the selected cross sections, aerodynamic coefficients are measured by the wind tunnel tests.
3. Static stability verification. With measured aerodynamic coefficients, static instabilities such as lateral buckling or divergence are checked.
4. Dynamic verification. Dynamic phenomena such as self-excited vibration, gust response, and vortex-induced oscillation are verified. The self-excited vibration is a phenomenon in which amplitude of vibration is rapidly amplified by the provision of aerodynamic force generated by the movement of the structure itself. Galloping and flutter are forms of self-excited vibration. Although this kind of dynamic instability was once verified only by the wind tunnel tests, today, numerical analysis is conducted with measured unsteady aerodynamic coefficients. The gust response is the dynamic response induced by the fluctuation of wind speed. It is verified by the wind tunnel tests or numerical analysis based on the random response theory. The vortex-induced oscillation is a limited-amplitude resonance caused by the vortices generated in the wake of the structure itself and observed in relatively low wind velocity. Types of aerodynamic phenomena are shown in Figure 2.10.

2.3.5 Seismic Design

Since the suspension bridge has a longer natural period than other types of structures in general, seismic design may not be dominant in the design. If the suspension bridge is built in an earthquake-prone area, adequate

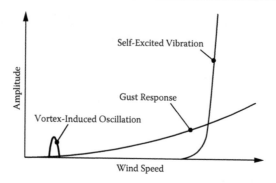

Figure 2.10 Types of aerodynamic phenomena.

seismic design must be conducted. The response to earthquakes is calculated by using response spectrum analysis or time–history analysis.

For the seismic design of long-span suspension bridges, seismic input having long period components should be considered because the frequency of these bridges is usually very low. Also, if the center span is very long, the time lag of the input seismic force for each foundation may affect the response of the bridge.

2.3.6 Main Towers

For the transverse direction, analysis based on the infinitesimal deformation theory was conducted because the restriction from the main cable is negligible, and the transverse stiffness of the main tower is very high. For the longitudinal direction, the method proposed by Birdsall (1942) has been broadly used. However, with the advancement of computer technology, three-dimensional models with finite deformation theory are generally used to calculate for both directions.

2.3.7 Main Cables

Alignment of the main cable is controlled by sag of the center span. Since the total cost of the suspension bridge varies with the sag–span ratio, it is common to select the ratio that gives the most economical solution in the preliminary design. However, since the sag also affects the stiffness and dynamic characteristic of the whole suspension bridge, it is important to verify the aerodynamic stability. After deciding the sag in the center span, sags of the side spans are to be decided with the equilibrium of horizontal cable tension at the top of the main towers.

2.3.8 Suspended Structures

The width of the stiffening girder is designed to accommodate traffic lanes and adequate shoulders. On the other hand, height is occasionally decided

Table 2.1 Characteristics of Stiffening Girders

	Truss Girder	Box Girder
Height	High	Low
Aerodynamic stability	Self-excited vibration shall be verified	Vortex-induced oscillation tends to occur Self-excited vibration shall be verified
Maintenance	Coating area is large	Coating area is small
Construction	Cantilevering method or large-block-hoisting method	Hoisting method

to secure aerodynamic stability since it affects the bending and torsional stiffness of the bridge. The aerodynamic stability of the stiffening girder is verified by the wind tunnel tests after the preliminary design.

For general suspension bridges, the type of the stiffening girder is either truss girder or box girder. Table 2.1 shows the characteristics of the two types of stiffening girders to be considered for selection.

Except for verification of the aerodynamic stability, the cross section of the stiffening girder is mainly decided by live load or static wind load. For vertical planes, linearized finite deformation analysis is conducted because it requires influence line analysis for live load. For wind load, static wind force is applied to the three-dimensional frame model. Linearized finite deformation analysis is conducted since the variation of cable tension is considered to be small in the out-of-plane analysis.

2.4 CONSTRUCTION

2.4.1 Construction Sequence

Figure 2.11 shows a construction sequence of a general suspension bridge. Construction starts from the anchorages and the main towers. Then the main cables are spun between the anchorages via main towers, and the stiffening girders are hung from the main cables. The installation of the stiffening girder can start from either the center of the center span or the main towers. The former method is advantageous since it can reduce the secondary stress in the cable, which is induced when the cable bands are installed, near the main towers because the cable is almost final configuration when the final portion of the stiffening girder is installed at the towers. The latter also has the advantage of allowing workers to use the installed portions as scaffolds.

2.4.2 Main Towers

For the steel main tower, components are fabricated in the factories and shipped to the construction site. Each component is erected by cranes, piece by piece. For the concrete main tower, slip forming is normally used.

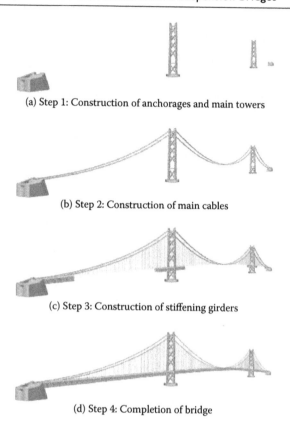

(a) Step 1: Construction of anchorages and main towers

(b) Step 2: Construction of main cables

(c) Step 3: Construction of stiffening girders

(d) Step 4: Completion of bridge

Figure 2.11 Construction sequence of suspension bridge.

Since the main tower is designed to be stable in the final configuration, that is, with main cables and suspended structures, it may require counter-measures to suppress oscillation during the construction stage. Before the installation of the main cables, the main tower is considered to be a can-tilever without any restriction at the tower top. As a countermeasure, the freestanding tower is, in some cases, stiffened with the supporting cables. For the modern suspension bridge, vibration control devices such as tuned mass dampers (TMDs) are installed. Photo 2.1 shows the TMD used in the Akashi Kaikyo Bridge. These TMDs were left in place even after the com-pletion of the bridge to reduce the amplitude of vortex-induced vibration.

2.4.3 Main Cables

Erection of the main cables starts from the construction of scaffolding, called catwalks. To construct catwalks, a pilot rope is spun between the anchorages via main towers. There are various spinning methods for the pilot rope. Although ships or floating-crane ships are traditionally used,

Photo 2.1 Tuned mass damper for Akashi Kaikyo Bridge.

a helicopter can be used for modern suspension bridges that span busy marine traffic lanes. Once the catwalk is constructed, cable components are built up. As mentioned in Section 2.2.4, there are two types of cable construction methods for the parallel wire cable.

The AS method has been conventionally used for over 100 years, being considered an on-site method because the cable is actually built up from single wires delivered in coils. The total length of the wire in one coil reaches 10 to 20 times the distance between the anchorages. The wires are drawn by the spinning wheel from one anchorage to another. When the spinning wheel reaches the anchorage on the other side, the wire is set onto the strand shoe and the wheel goes back to the original anchorage and repeats the same sequence (Figure 2.12). Since the sag is adjusted during air spinning, this method is affected by the wind condition. This conventional AS method requires a significant number of labor workers to minimize the variability of wire length. The improved AS method controls the wire tension during the spinning to obtain uniform wire length, and therefore does not require adjustment of individual wires.

The prefabricated parallel wire strand (PPWS) method was developed to reduce the construction time needed for the AS method. In this method, prefabricated strands composed of about 100 wires and with a length equal to the cable length between the anchorages are used. The sockets are set at both ends of the strand (Figure 2.13). Each strand is drawn from one anchorage to the other using hauling rope.

2.4.4 Suspended Structures

The prefabricated unit with a larger size is advantageous in the construction of the stiffening girders since it can reduce the construction time and keep high quality. Therefore, the large-block-hoisting method, where the

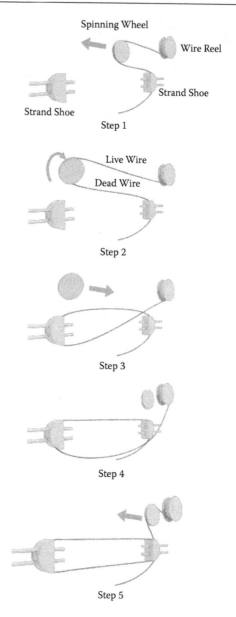

Figure 2.12 Concept of air-spinning method.

large prefabricated girder blocks are shipped by the barge to the site and hoisted up by the hoisting equipment installed on the main cables, is preferred if the water surface is available just beneath the bridge. Even if the water surface is not available, the blocks can be hoisted from the available area (usually the vicinity of the main tower, which is outside of the shipping

Figure 2.13 Prefabricated parallel wire strand.

lanes) and transferred longitudinally by suspender ropes; this is referred to as the swing method. The large block can be installed by the floating-crane ship, although interference of the crane boom to the suspenders must be considered. If the water surface is not available because of busy marine traffic or rapid tidal current, the cantilevering method is used. The smaller members are installed piece by piece from the cranes set at the tips of the stiffening girders. Photos 2.2 and 2.3 shows examples of the large-block-hoisting method and cantilevering method, respectively.

Photo 2.2 Large-block-hoisting method (Kurushima Kaikyo Bridge).

Photo 2.3 Cantilevering method (Akashi Kaikyo Bridge).

Before the completion of the stiffening girder, the girder blocks are either rigidly connected or hinged. Hinged connection is preferable since the final connection of the girder blocks is easy after the installation of all the girder blocks, and it does not induce the excessive tension force to the suspenders. However, the aerodynamic stability during construction is poorer than the rigid connection method, and therefore it requires certain countermeasures against the strong wind.

REFERENCES

Birdsall, B. 1942. The suspension bridge tower cantilever problem, 847–862. *Trans. ASCE*, Vol. 107.

Honshu-Shikoku Bridge Authority. 1990. *Wind-Resistant Design Standard for the Akashi Kaikyo Bridge*. Honshu-Shikoku Bridge Authority.

Melan, J. 1888. Theorie der eisernen Bogenbrücken und der Hängebrücken. In *Handbuch der Ingenieurwissenschaften*. Leipzig, Germany.

Moisseiff, L. S. and Lienhard, F. 1933. Suspension bridges under the action of lateral forces. *Trans. ASCE*, 58.

Nader, M., Baker, G., Duxbury, J., and Maroney, B. 2013. Fabrication and construction of self anchored San Francisco Oakland Bay Suspension Bridge. In *Durability of Bridge Structures*, ed. Mahmoud, 17–29. Boca Raton, FL: CRC Press.

Navier, M. 1823. *Mémoire sur les Ponts Suspendus*. Imprimérie Royale, Paris, France.

Chapter 3

Inspection of Suspension Bridges

Justine Lorentzson, Yimin Chen, Lloyd Hansen,
Peter McDonagh, Jaroslaw Myszczynski,
and Dora Paskova

CONTENTS

3.1 INTRODUCTION

Suspension bridges are graceful and aesthetic. More importantly, they are efficient and economical structures for the lengths they span. The basic concept is not new: China reportedly used this concept as early as 200 BC. The development of the suspension bridge concept has been steady, and there are numerous examples that can be cited. Improvements in materials led to the first chain suspension bridge in Europe in 1741 (Winch Bridge, near Bowleed in UK). Later, an improved version was constructed in the United States, supported by iron chains similar in profile to today's bridges, which remained in service from 1801 to 1833 (Jacobs Creek Bridge, just south of Mount Pleasant, Pennsylvania). Continued improvements led to the use of wire cables and deployment of suspension bridges over larger spans. The Wheeling Suspension Bridge in Ohio (1010 ft tower to tower) was constructed in 1849 and is still in service. In 1854, a windstorm destroyed the deck through torsional and vertical movements. The deck was rebuilt, and subsequently, stay cables were added in 1871. Aerodynamic issues were not fully understood at that time, as became evident later with the collapse of the Tacoma Narrows Bridge in 1940. The Bronx-Whitestone Bridge was, like the Tacoma Narrows Bridge, built with a slender profile susceptible to aerodynamic problems. Improvements to the behavior of the bridge in high winds were achieved with the installation of a stiffening truss in 1941 and a tuned mass damper in the mid-1980s. Further improvements were achieved with the replacement of the bridge deck, relocation of the bridge lateral bracing, and installation of fairing. With every new suspension bridge constructed, the lessons learned from previous bridge performances have been incorporated, leading to an ever-improving design for suspension bridges. At present, the Akashi Kaikyo Bridge, which connects the city of Kobe on the mainland of Honshu to Iwaya on Awaji Island in Japan, is considered the longest suspension bridge in the world, with a central span of 1991 m.

3.2 INSPECTION OF SUSPENSION BRIDGES

Suspension bridges are an attractive choice due to a number of factors: the long distances they cover, the limited substructure required, potential savings in material costs, and the fact that the main suspension cables essentially function in tension only. However, suspension bridges are still large investments and require considerable effort in terms of monitoring and maintenance. Failures have accentuated the point: the failure of the Silver Bridge's eyebar chain in 1967 led to the development of the current bridge inspection standards in the United States and the rest of the world. Currently, all bridges in the United States are required to undergo complete inspection, typically on a 2-year cycle as mandated by the National Bridge Inspection Standards governing highway bridges [1]. Some bridges are eligible for inspection on a 4-year cycle, but this rule does not apply to suspension bridges because of their length, lack of load path redundancy, and significant number of fracture-critical members [1, 2]. In addition, some states (including New York State) do not allow an extended inspection cycle and mandate that all highway bridges be inspected on a biennial basis, with more frequent inspections when warranted [2]. The content of this chapter is based on the authors' experience in managing suspension bridges in New York City, but in general applies to most suspension bridges in United States and the rest of the world.

During a regular biennial inspection, all bridge components are visually inspected. This level of inspection is adequate for a number of components, including the following:

- Deck and deck joints
- Structural steel supporting the roadway
- Stiffening elements (truss, box or I-girders, etc.)
- Hand ropes and supporting stanchions
- Guide rails
- Substructures

For some suspension bridge components (nonredundant, fracture-critical members), the visual inspection performed under the routine biennial inspection only provides a minimum level of inspection. These bridge components require additional, special inspections between normal inspection cycles to ensure that they continue to function as originally designed [4]. For example, cables, eyebars, anchorages, and towers require this level of inspection. Detailed inspections of suspension span towers, considered nonredundant bridge components, are required since the towers support the cable vertical loads and resist horizontal forces from the spans and the cables. Likewise, the anchorages are equally important, as they resist the load of the main cables. Further discussion of critical suspension bridge components is provided in the following sections, which identify additional inspection and testing needed to fully track their condition and ensure the structural integrity of the bridge.

Design limitations and original material issues play an important role in the suspension bridge longevity. Factors that affect longevity, such as maintenance shortcomings, fatigue damage, and material embrittlement, must be identified via in-depth inspections. As suspension bridges have aged, inspection and testing of these critical elements have become more common, with every generation adding to the list of inspection methods. In addition, published guidelines have also evolved with the inspection of critical suspension bridge components. Industry guidelines and the standard industry practices regarding the inspection of critical suspension bridge members will be the focus of the remainder of this chapter. Published industry standards and guidelines for inspection of suspension bridges in the United States include the following:

- New York State Department of Transportation (NYSDOT) *Bridge Inspection Manual* (BIM) [3]
- *Bridge Inspector's Reference Manual* (BIRM), where Section 12.1 covers inspection of cable-supported bridges [5]
- National Cooperative Highway Research Program (NCHRP) Report 534: *Guidelines for Inspection and Strength Evaluation of Suspension Bridge Parallel Wire Cables* [6]
- *Primer for the Inspection and Strength Evaluation of Suspension Bridge Cables* (FHWA-IF-11-045) [7]

3.3 CRITICAL SUSPENSION BRIDGE ELEMENTS

3.3.1 Towers and Associated Assemblies

The towers of a suspension bridge carry the vertical forces from the cables, with the tower saddles transferring the load from the cables to the tower structures. The towers are a nonredundant feature of the suspension bridge. Steel towers, in particular, are vulnerable to corrosion from atmospheric exposure and water infiltration, and chemical attack from chlorides, nitrates, and ammonia and acids resulting from bird droppings, which can compromise various components of the towers. It is essential to identify any means of water ingress, such as holes in the exterior plates, open hatches, and so forth, which not only let in water, but also birds. During the biennial inspection, a visual inspection of the tower steel and all interior and associated tower components is performed. Paint conditions are noted, as well as the condition of the tower steel, along with signs of water and bird infiltration.

There are several key elements associated with the towers that require special attention and inspection during the biennial inspection, and a few of these elements require special attention beyond the biennial inspection. These elements, including the tower saddles, hangers, wind tongues, finger joints, and drainage system, are further discussed below.

3.3.1.1 Tower Saddles

The tower saddles are usually cast steel elements that cradle the main cable as it passes over the tower, transferring the load from the cables into the towers, allowing for a gradual change of cable direction from one side of the tower to the other, preventing lateral movement of the cables (Figure 3.1).

Biennial hands-on inspection is recommended for the tower saddles. Such inspections should identify any debris accumulation on the cable within the saddle and among the stiffener ribs leading to moisture retention and

Figure 3.1 Tower saddles.

corrosion of bolts and support frame members. The condition of the paint system should also be documented.

3.3.1.2 Hangers

The hangers (Figure 3.2) are steel members designed to limit vertical movement while allowing longitudinal movement of the deck at the towers. They are not designed to resist lateral movement. The vertical loads in the hangers are transferred through pin and bushing connections at each end

Figure 3.2 Tower hanger.

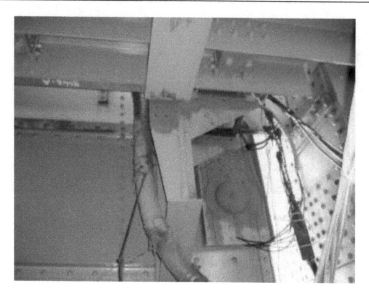

Figure 3.3 Hanger pin (inside a girder).

(Figure 3.3). These pin caps should be removed for inspection on a 4- to 6-year cycle. Regular cleaning and lubrication are also recommended.

A biennial hands-on inspection is recommended for the hangers to identify any signs of corrosion, visible deformation or warping, cracking of the steel, pin eccentricity, and the condition of the paint. The hangers should also be inspected after every major wind or seismic event. When there are indications of cracking, dye-penetrant testing is utilized to determine the extent of the cracking.

3.3.1.3 Wind Tongues

The wind tongues are elements that allow longitudinal movement of the deck at the tower and actively prevent lateral movement, such as those caused by wind loading. Essentially, the wind tongue is a reinforced steel box built into the lower strut outrigger with the longitudinal dimension larger than the deck movements caused by thermal expansion and contraction, or live load. There is a rectangular deck wind pin tightly fitted inside this box. The inner longitudinal surfaces of the outrigger box are lined with perforated 1 in. thick bronze plates, while the outer longitudinal surfaces of the deck wind pin are lined with solid 1 in. thick bronze plates. The perforations in the outrigger box liners allow extrusion of grease pumped in through the grease fittings on each side of the wind tongue. The arrangement is designed to limit friction between the liner plates while allowing longitudinal motion of the pin inside the outrigger box. Failure of the wind tongues could lead to deformation of the hangers and structural damage to the towers and decks.

There are two levels of inspection of the wind tongues that are recommended:

1. Routine inspection by maintenance or engineering personnel: The inspection focuses on accumulation of debris or changes in the appearance of the grease. It is recommended that these inspections be performed annually and after any major wind or seismic event.
2. Biennial hands-on inspections: Biennial inspections in New York State require that nonredundant members receive hands-on inspection. For the wind tongues, this includes visual inspection of the condition of grease and visible deformation or cracking of steel. In addition, the biennial inspections should monitor the wearing of the bronze plates with a caliper gauge. Uneven wearing may occur depending on the prevailing wind direction, and thus is usually more pronounced on the downwind side.

3.3.1.4 Finger Joints

The finger joints (Figure 3.4) are expansion joints, that is, steel closure elements between the side span and main span decks allowing independent longitudinal movement of each deck while providing support to the vehicular traffic on the deck. Depending on the roadway width, a finger joint may consist of several assemblies extending to multiple fingers. The number of fingers on each assembly varies depending on the design, with each assembly weighing several hundred pounds. The assemblies are mounted to the deck structure with springs, either steel or heavy-duty neoprene, allowing limited deflection. In the center of the space between the side and main spans, there is a transverse support beam with a convex top surface allowing the fingers to slide over it. The finger joints and support beams (Figure 3.5) are part of

Figure 3.4 Typical finger joint at a tower.

Figure 3.5 Support beam under the finger joint.

a system, with a drainage trough underneath, to prevent the surface runoff from draining through the finger joint directly onto the lower strut.

Since the finger joints sustain significant impact forces, they require frequent inspection to quickly identify issues such as missing fingers, excessive vertical movement under wheel loading, or clanging noise when vehicles pass over the joint. Vertical movement or clanging noises are indicative of problems such as cracks in the joint assemblies, loose bolts, misalignment between adjacent assemblies, cracked, loose, or missing springs, and missing nuts, which, if left unrepaired, can lead to catastrophic failure of the joint with potentially serious safety issues for the traveling public.

There are two levels of inspection of the finger joints that are recommended:

1. Routine inspection by maintenance or engineering personnel: Weekly visual inspections from the top of the deck are recommended. Personnel should look for missing fingers, excessive vertical movement under wheel loading, clanging noise when vehicles pass over the joint, cracks in the joint assemblies, or misalignment between adjacent assemblies. If any of these issues are noted, a more detailed hands-on inspection should be conducted immediately for an in-depth inspection of the joint from underneath to determine the extent of the problem.
2. Biennial hands-on inspections: These include visual inspections of the top and underside of the finger joints and all parts of the finger joint assembly.

3.3.1.5 Modular Joints

In lieu of finger joints, some suspension bridges have modular joints at the towers. The modular joints allow greater movements than finger joints through the use of multiple elastomeric seals that form a modular bridge joint system. A modular joint includes one or more transverse center beams parallel to the edge beams to separate two or more elastomeric seals, thus increasing the total possible movement. Similar to finger joints, modular joints require very specific inspection and maintenance.

3.3.1.6 Drainage

Proper drainage within steel towers is very important to make sure that there is no ponding of water in the cells, leading to corrosion of the steel. Accumulation of debris can block the drain holes and prevent proper drainage. A hands-on inspection every 2 years is recommended unless more frequent inspections are warranted based on the observed conditions by maintenance and engineering personnel. In addition, the towers should be inspected for water retention after major storms with high winds driving rain at angles, to ensure that there is no accumulation of debris and the weep holes are free to drain.

3.3.2 Main Cables/Suspender Ropes

The main cables of a suspension bridge carry the dead load from the deck and the imposed live load and transfer these loads to the towers. The cables act in tension and require anchoring at both ends into either massive concrete blocks or rock. The main cables are nonredundant members of the suspension bridge. The suspender ropes loop around the main cable at each cable band and are socketed into a girder or truss system that supports the deck. These vertical cables transfer the loads from the girder or truss to the main cables.

During the biennial inspection, a visual inspection of the suspender ropes and cables is performed by walking the cables and looking for signs of deterioration, such as rust, deteriorated paint conditions, and so forth. The suspender ropes and sockets are also inspected for any visual signs of distress/deterioration. There are several key components of the main cable and suspender rope system that require additional inspection beyond the biennial inspections. These elements—the main cables, the cable bands, and the suspender ropes and sockets—are further discussed below.

3.3.2.1 Main Cables

The main cables are constructed of many individual wires, typically with a 0.192 in. diameter and a zinc coating that protects the wires from corrosion. The wires are compressed and secured typically every 40 ft or so with a cable band, around which the suspender ropes also hang. In addition, the compressed cables are often coated with an anticorrosive coating, such as red lead paste or zinc paste, then wrapped with wrapping wire, and painted to keep water out of the cables. Where the wrapping wire meets the cable bands, caulking is applied to complete the water-resistant seal. In addition, some cables have dry air dehumidification systems installed as an added protection against moisture infiltration. The main cause of deterioration of main cables is water infiltration causing rusting and hydrogen embrittlement of the cable wires. It is impossible to accurately assess the internal condition of the main cables via visual inspection without opening the cable; however, visual inspection does help with ensuring the cable protection system (i.e., wrapping wire and coatings/caulking) is maintained in good condition.

There are three levels of inspection of the main cable that are recommended:

1. Periodic routine visual inspections by maintenance or engineering personnel of the cable exterior: The inspection focuses on changes in the appearance of the cable—damage to the paint or wrapping caused by accidents, weathering of the paint system, loose caulking, and so forth. It is recommended that these inspections be performed at the end of winter and summer months.

2. Biennial hands-on inspections: Biennial inspections require that nonredundant members receive hands-on inspection. For the cables in suspension spans, this includes reporting and rating the condition of the paint, caulking at cable bands, wire wrapping, and bottom of cable or cable bands. For the cables inside anchorages, this includes reporting and rating corrosion or broken wires of the strands, swelling or bulges at the strand shoes, water entry at the anchorage walls and roof, condensation and corrosion of eyebars and strand wires, and corrosion at the contact surfaces between eyebars and the concrete mass.

3. Internal inspections: The goal of the internal suspension bridge cable inspection (Figure 3.6) is to obtain information about its condition and strength. This is accomplished by taking wire samples from within the cable and performing destructive testing on the wires to determine their remaining strength. A visual inspection of a portion of the individual wires that make up the cable is also conducted. Internal inspections are recommended for all cables during their lifetime, regardless of the cable's external appearance.

The internal cable inspection consists of the following activities:

- Selection of panels to be inspected
- Unwrapping and wedging of the cable
- Visual inspection of the visible wires (Figure 3.7)
- Wire sampling and splicing
- Compacting and rewrapping of the cable

Figure 3.6 Unwrapped section of main cable, showing inspection wedging. Broken wires are readily identifiable.

Figure 3.7 View into a wedged open cable showing varying degrees of wire deterioration.

The locations of the internal inspections are based on results of past inspections, acoustic monitoring data when available (acoustic monitoring systems are only present on a few bridges), and external signs of internal deterioration. If no results or data are available, the locations of the first inspection can be selected based on the recommendations of Section 2.2.5.1 of NCHRP Report 534 [6]. The internal inspection intervals suggested by NCHRP Report 534 [7] are presented in Table 3.1. At the discretion of

Table 3.1 Suggested Internal Inspection Intervals Based on NCHRP Report 534

Inspection Number	Maximum Corrosion Stage Found in Previous Inspection	Age of Bridge at Last Inspection (Years)	Interval (Years)
First			30
Additional	1-(2)	Any age	30
	2-(3)	40 or more	20
	2-(3)	30	10
	3-(4)	60 or more	20
	3-(4)	Less than 60	10
	4	Any age	10
	Broken wires	Any age	5

Source: National Cooperative Highway Research Program, *Guidelines for Inspection and Strength Evaluation of Suspension Bridge Parallel Wire Cables*, Report 534, National Cooperative Highway Research Program, Washington, DC, 2004.

Note: Each corrosion stage may include up to 25% of the surface layer of wires in the next higher stage, indicated by the number in parentheses. Stage 4 may include 5 broken surface layer wires.

the owner and the investigator, the inspection intervals could be adjusted. However, they should not exceed 5 years when stage 4 corrosion is found in more than 10% of the wires in the cable.

At each location, the existing wrapping wire should be removed and examined. The next step is to clean and wedge the cable. NCHRP Report 534 provides detailed instructions on the process of wedging—incrementally and sequentially at eight circumferential positions. Traditionally, wedging starts from the bottom and progresses to the upper quadrants. This sequence is used to find the suspected worst-condition wires first, along the bottom of the cable, where water tends to collect and initiate corrosion. When wedging, caution must be taken not to damage cable wires and zinc coating, and to avoid any crossing of wires along the groove. Upon the completion of the wedging, a detailed inspection of surface and interior wires should be performed consisting of visual inspection for the full length of the groove and detailed photography along each cable.

Table 3.2 provides the number of samples of different corrosion stage wires recommended for laboratory testing as per NCHRP Report 534 [6]. The goal is to remove the minimum number of wires for laboratory testing that would provide a reliable cable strength estimate. It is important that the results from prior inspections be taken into consideration when determining the number of wires to be removed for laboratory testing. Specifically, once the total number of removed and tested stage 1 and stage 2 wires equals the recommended number of wires, no further samples of these stages are required in further inspections.

Table 3.2 Recommended Wire Samples for Laboratory Testing

Corrosion Stages Present in Worst Panel Observed				Total Number of Samples				Estimated Error (97.5% confidence) Bridge		Estimated Cable Strength Loss Bridge	
Stage 1	Stage 2	Stage 3	Stage 4	Stage 1	Stage 2	Stage 3	Stage 4	X	Z	X	Z
100%	0%	0%	0%	10	—	—	—	3%	5%	0	0
	>0%	0%	0%	10	15	—	—	3%	5%	0%	0%
	>0%	10%	0%	10	15	35	—	3%	5%	1%	2%
	>0%	20%	10%	10	15	35	60	4%	5%	9%	10%
	>0%	40%	20%	10	15	35	60	4%	6%	16%	18%

Source: National Cooperative Highway Research Program, Guidelines for Inspection and Strength Evaluation of Suspension Bridge Parallel Wire Cables, Report 534, National Cooperative Highway Research Program, Washington, DC, 2004.

Laboratory testing of the wire samples removed during internal inspection is an integral part of the cable inspection. The following tests provide information on the strength and condition of the wires and the cables: visual inspection, strength tests, fractographic evaluation, chemical analysis, hydrogen content, and surface corrosion analysis.

3.3.2.2 Alternative Methodology for Cable Inspection/Strength Modeling

An alternative methodology to the ones provided in the National Cooperative Highway Research Program (NCHRP) Report 534 is the Bridge Technology Consulting (BTC) method. The BTC method includes random wire sampling without regard to wire appearance, mechanical testing of wire samples, and the calculation of the probability of broken and cracked wires. Additional information on the BTC method and its applications is available in the *Primer for the Inspection and Strength Evaluation of Suspension Bridge Cables* developed for the Federal Highway Administration by HDR Engineering [7].

3.3.2.3 Suspender Ropes/Suspender Rope Sockets

The suspender ropes are constructed of many individual wires and are typically 2¼–2½ in. in diameter. They consist of a core strand of wires that is surrounded by braided strands, and at each end of the suspender rope there is a large zinc socket. Experience shows that other than a few surface broken wires, deterioration of the suspender ropes (Figures 3.8 through 3.11) is very difficult to spot during visual inspections since the majority of the deterioration occurs on the center strand of the rope, which, due to the construction of the rope, is completely hidden from view by the outer braided strands. In addition, significant deterioration has been noted at the interface of the rope with the socket, as well as in the area where the rope passes through the girder.

Figure 3.8 Typical bleeding of rust near bottom of suspender rope wires.

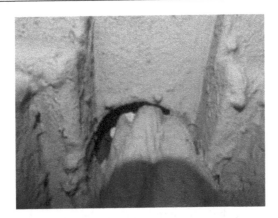

Figure 3.9 Suspender rope offset from center of hole through girder.

Figure 3.10 Typical broken wire on suspender rope and girder.

Figure 3.11 Typical pack rust found between suspender ropes.

There are three levels of inspection of the suspender ropes that are recommended:

1. Biennial hands-on inspections: As part of the biennial inspection, the suspender ropes are visually inspected, including the socket areas. The condition of the paint is rated and reported along with any observed broken wires both in the main body of the ropes and at the socket interface.
2. Nondestructive inspection/testing of straight portion of the ropes: This type of inspection utilizes a technology based on the principles of magnetostriction and is used to identify reduction in cross section area due to corrosion in the straight sections of the rope. This method cannot access and obtain readings in the area where the suspender ropes pass over the cable bands, nor at the socketed ends. Depending upon the age/condition of the suspender ropes, it is recommended to perform this testing on all of the ropes at least once every 5 years, along with the main cable inspection.
3. Destructive testing of ropes: Both straight-pull tensile tests (Figure 3.12) and tensile tests over the sheave (Figure 3.13) are performed. The goal of the destructive testing of the suspender ropes is to determine the remaining strength of the ropes and socketed ends and to determine the internal condition of the rope being tested. Random socket sectioning (Figure 3.14) is also performed to determine defects in the sockets. Ropes for destructive testing can be selected based upon results from

Figure 3.12 Straight-pull tensile test (see ASTM A586 and A370).

Figure 3.13 Tensile test over sheave (see ASTM A586 and A370).

Figure 3.14 Socket sectioning.

nondestructive testing, visual inspection, or random samples. For older bridges, it is recommended to perform destructive suspender rope testing when an in-depth inspection of the main cable is performed.

3.3.2.4 Cable Bands/Cable Band Bolts

Cable bands (Figure 3.15) are two halves of a cylinder bolted together over the circumference of the cable, typically at even spacing dictated by the suspender rope spacing. The cable bands provide a seat for the suspender ropes, transferring the load from the ropes to the main cable. The friction forces between the cable band and the cable wires keep the cable band from sliding down the cable under these forces. The bolts on the cable bands have a tendency to loosen over time, and need to be regularly inspected to ensure that proper tension is maintained in the cable band bolts. It is recommended that a portion of the cable band bolts be tested for torque every time a cable inspection is performed.

It should be noted that while torque can give a good indication of bolt tension, it can vary with the level of corrosion at the threads. Highly variable conditions of corrosion can be confirmed with an extensometer, measuring bolt length at rest and after tensioning. At a minimum, the cable band bolts on either side of the cable panels that are opened for inspection should be inspected and tested. Depending on the findings of the inspection, it is recommended to inspect up to 20% of the cable band bolts once every 5 years.

3.3.2.5 Anchorages

The main cables terminate in the anchorages, where each cable enters through a collar, passes over a cable saddle, and splits into individual strands

Figure 3.15 View of a typical cable band and bolts.

anchored into the concrete mass via strand shoes and eyebars. The cable saddle is a steel casting on which the cable rests. It resists the vertical loads imposed by the cable as the cable is redirected downward toward the anchoring mass. The cable saddle also controls the flare of the cable strands as the cable separates into individual strands that anchor into the eyebars. The concrete anchorages provide the mass necessary to resist the uplift forces from the main cables and provide protection from the elements for the cable anchoring devices, including strand shoes, cable strands, eyebars, and cable splay saddles. Often, the anchorages also support the roadway deck transitioning from the approach spans to the suspended spans.

The concrete anchorage mass is inspected per the routine biennial inspection and after extreme events such as earthquakes or hurricanes. The biennial inspection of the anchorage elements is conducted in accordance with applicable bridge inspection standards [3, 5]. The cables strands, cable strand shoes, and eyebars are fracture-critical members that receive 100% hands-on inspection during the biennial inspections. Particular attention is paid to the cable strands at the point of tangency with the strand shoes and the eyebars at the interface with the concrete anchorage block. There are several key elements in the anchorages that require special attention and inspection beyond the biennial inspection. These elements—the cable strands, cable strand shoes, eyebars, cable strand enclosure and dehumidification systems, and anchorage drainage system/anchorage interior concrete—are further discussed below. The elements of the main suspension system inside the anchorages, including the anchorage structure, and equipment that are part of the inspections are also further discussed below, along with additional recommended inspections.

3.3.2.6 Cable Strands

Unlike the main cable outside of the anchorage, the cable strands inside of the anchorage are not wrapped; however, in some cases the strands have been painted. The main issue with the cable strands is corrosion caused by the damp and humid interior environment or pigeon droppings exacerbated by water infiltration. Like the main cable, it is impossible to accurately assess the internal condition of the strands through visual inspection; however, visual inspection does help with identifying surface corrosion progression and broken wires (Figure 3.16). The strands inside the anchorage should be wedged for more detailed inspection. Experience shows that special attention should be given to the bottom half-strands, near and at the contact of the strands with the strand shoes.

Depending upon the bridge, deterioration of the strands where they wrap around the strand shoes has been identified as a serious issue (Figure 3.17), leading to the reanchoring of the affected strands (Figure 3.18).

There are two levels of inspection of the cable strands that are recommended:

Figure 3.16 Typical cable strands inside an anchorage with a visible broken wire.

Figure 3.17 Excessive corrosion of strand where it wraps around the strand shoe.

Figure 3.18 Reanchored strands.

1. Biennial hands-on inspections: Biennial inspections [1–3] require that nonredundant members receive 100% hands-on inspection. For the cable strands in the anchorages, this includes reporting the presence of broken wires and corrosion on the exterior of the strands. In addition to the federally mandated biennial inspection requirements, owners such as the MTA Bridges and Tunnels have additional technical requirements for cable strand inspection, which call for a minimum of 10% of the cable strands to be wedged open for inspection. The selected cable strands are wedged where there is visible corrosion on the exterior wires of the strand. This wedging procedure to open the selected strands involves inserting two plastic wedges into or through the strand using a hammer to open the strand over a 4.0 to 6.0 ft length (Figure 3.19). The interior wires are inspected with a magnifying glass for conditions such as the effectiveness of protective oil coating (if such a coating has been applied), water infiltration, zinc or ferrous corrosion (categorized as stage 1 to 4 wire corrosion per NCHRP Report 534 [6]), broken wires, pitting, wire section loss, cleanliness, and the presence of foreign matter. All such conditions are noted by strand number, location along the strand, number of affected wires, and depth into the strand interior. The affected cable strands are also tagged for future reference.

2. Full internal inspections: The goal of the internal strand inspection is to obtain information about the internal condition of all strands, and determine the need for additional protective oiling of the strands, if oiling is used for protection. Broken wires are spliced and repaired if observed. Full wedging of all strands is performed, and the condition rating of the exposed wires is noted. Depending upon the internal conditions noted, it is recommended that the full internal inspection

Figure 3.19 Typical wedging and inspection of cable strands within the anchorage.

of all cable strands be performed once every 5 to 10 years. On bridges with excessive corrosion where the strands wrap around the strand shoes, it is also recommended that those strands be jacked away from the strand shoes for more detailed inspection of the condition of the strands in that area.

3.3.2.7 Strand Shoes and Eyebars

The strand shoes are the elements the cable strands wrap around. The strand shoes are attached to the eyebars via a nut-and-bolt connection (Figure 3.21), and the eyebars are embedded into the concrete mass of the anchorage, resisting the massive tensile forces imparted by the main cable. The eyebars are tied together at the top to help resist the bending forces exerted due to the splay of the cable strands.

Typical issues experienced with these elements are corrosion of the eyebars where they interface with the concrete mass, resulting in significant reduction of the cross section (shown in Figure 3.20). The corrosion is always due to either high-humidity environments or water infiltration into the anchorages. Another issue that has been encountered on some bridges is pack rust between the tie plate, eyebar, and strand shoe and the rusting of the eyebar bolts that connect the eyebars to the strand shoes (Figures 3.22 through 3.25).

There are two levels of inspection of the eyebars/strand shoes that are recommended:

1. Biennial hands-on inspections: Biennial inspections [1–3] require that nonredundant members, including the eyebars and strand shoes, receive 100% hands-on inspection. The presence of corrosion is noted, along with any issues, such as corroded nuts, and so forth.

Figure 3.20 View of eyebar–concrete interface with obvious necking down due to corrosion.

Figure 3.21 View of typical eyebar–strand shoe connection with tie plates.

Figure 3.22 View of typical tie plate–eyebar–strand shoe connection with deteriorated bolt–nut.

2. Full cleaning and inspection of eyebar–concrete interface: On many bridges the eyebar–concrete interface is protected with a coating of grease. This grease needs to be completely removed and the eyebars inspected for section loss. If there is a history of lack of maintenance and eyebar deterioration is present, it is recommended that the eyebars be fully inspected and measured for section loss once every 5 years. If the eyebar–concrete interface is in good condition, the full inspection should be performed once every 10 years, unless there is a working anchorage dehumidification system and there is no visual evidence of

Figure 3.23 Typical view of eyebar with section loss.

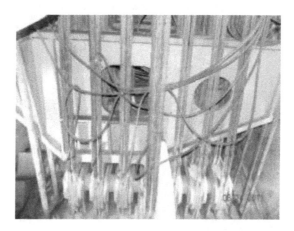

Figure 3.24 Excessive section loss can lead to the need to install a load-bearing system to reduce the load on eyebars.

deterioration, in which case the full inspection/measurement may be deferred indefinitely.

3. For critical members such as eyebars and reanchored strands, MTA Bridges and Tunnels has used x-ray diffraction, a nondestructive method for the evaluation of the remaining strength in eyebars and to determine the load carried by reanchored strands.

3.3.2.8 Anchorage Dehumidification System

In many anchorages, dehumidification chambers have been installed to isolate the eyebars and strands from the rest of the anchorage interior, in an effort to facilitate the dehumidification of these elements. It is generally

Figure 3.25 View of girder system installed at the Bronx-Whitestone Bridge to reduce the load carried by deteriorated eyebars.

accepted that if the relative humidity is less than 40%, active corrosion does not occur.

There are two levels of inspection of the anchorage dehumidification system that are recommended:

1. Monthly inspections of the dehumidification system by maintenance or engineering personnel are recommended to ensure the functioning of the system and to change filters: The relative humidity and temperature inside the chambers should also be measured and recorded during each inspection. A dusty environment will necessitate frequent filter changes. Dust that is not trapped by the filter clogs the desiccant wheel, causing system failure.
2. Biennial inspection: The anchorage dehumidification chamber is inspected for system integrity and equipment malfunction during the biennial inspection.

3.3.2.9 Water Infiltration/Drainage

Water infiltration into the anchorage and into the dehumidification chambers is a common issue, especially with older anchorages (Figure 3.26). In addition, clogging or malfunction of the anchorage drainage system is a common issue. Both issues, and especially the lack of proper drainage, can lead to excessive humidity, pooling of water, and active corrosion of critical anchorage elements.

There are two levels of inspection of the drainage system and anchorage concrete interior that are recommended:

Figure 3.26 Typical view of water infiltrating to the interior of the anchorages.

1. Monthly inspections, by maintenance or engineering personnel, of the interior of the dehumidification chamber drainage or the anchorage drainage system to ensure the drainage system is functioning and allowing any water that infiltrated the anchorage to drain out immediately: This inspection should be performed at the same time as the inspection of the dehumidification system.
2. Biennial inspection: The anchorage interior is inspected during the biennial inspection for signs of water infiltration and general dampness, and to identify any issues with the drainage system. Areas of water infiltration are noted for corrective action.

3.3.2.10 Paint Systems/Coatings

The paint systems and coatings used on suspension bridges are, in many cases, the first line of defense against deterioration of bridge components. Different paint/coating systems are used on each of the critical suspended span elements, and each has a different inspection protocol, as will be discussed below. The coating systems on the main cables, towers, suspender ropes, and eyebars are visually inspected as part of the routine biennial inspection. For the towers, this approach is sufficient, since the paint on the tower steel is the same paint used on the regular bridge steel, and is subject to the usual wear and tear that can be tracked through biennial inspections. The remaining elements require additional inspections to ensure that the coating systems are in fact performing their protective functions.

3.3.2.11 Main Cable Coatings

If maintaining a barrier to water ingress via paint and caulking is a primary concern, the following guidelines apply.

Suspension bridge main cables are generally coated with elastomeric acrylic-type coatings. These are heavy-duty water-based self-priming coatings that resist corrosion, rust, and moisture, while still retaining flexibily up to 200% elongation. Flexibility and elongation are important properties when coating main cables. A typical main cable-painting regime includes the coating system consisting of one spot prime coat and two full coats applied every 7–10 years. The application of caulk is used to seal and waterproof the cable bands. It is imperative that the caulking and coating remain intact to prevent water from infiltrating the cable. Typical defects seen in the main cable paint system are bulges, tears in the coating, loose or missing caulking, and so forth (Figures 3.27 through 3.29).

Figure 3.27 Typical bulging/perforation of paint.

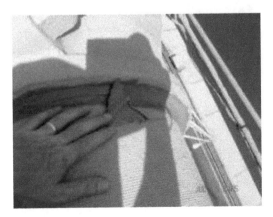

Figure 3.28 Typical damage to paint coating.

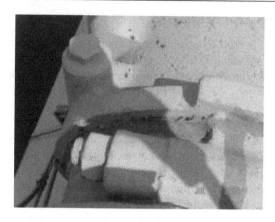

Figure 3.29 Typical debonding of caulk at cable bands.

There are two levels of inspection of the main cable coatings that are recommended:

1. Visual inspection by engineering staff every 6 months and after any work that is performed where the cables are used for access: A visual inspection of the main cables is performed by walking the main cables and using mirrors to view the bottom of the cables. Visual inspections should identify any bulging or tears in the paint, and missing or debonded caulking. These defects should be noted for immediate correction (Figure 3.30).
2. Biennial inspections: Visual inspections should identify any bulging or tears in the paint, and missing or debonded caulking. These defects should be noted for immediate correction.

Figure 3.30 Maintenance painting of the main cable.

3.3.2.12 Suspender Rope and Socket Coatings

Suspender rope and socket coatings usually consist of the same elastomeric acrylic-type coatings used for the main cable. Neoprene deflectors and caulking are often used where the suspender ropes pass through the truss or main girder. It is imperative that the deflectors, caulking, and coating remain intact to prevent water from infiltrating the suspender rope and causing rusting of the core strand. In addition, when water penetrates through the truss or girder, pack rust builds up above the socket and behind the rope, making it impossible to inspect the rope and socket in these areas. Typical defects seen in the suspender rope are tears in the coating, loose or missing caulking at the girder, and a buildup of pack rust between the rope and the truss or girder (Figures 3.31 through 3.33).

Since it is difficult to inspect the suspender ropes due to access issues, and since suspender ropes have more redundancy than the main cable, it

Figure 3.31 Typical missing paint on a suspender.

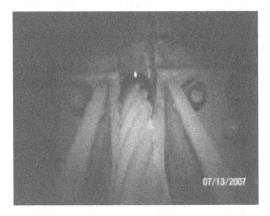

Figure 3.32 Typical gap in caulking, allowing water penetration to the socket area.

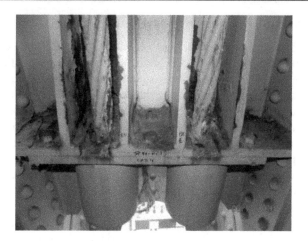

Figure 3.33 Typical pack rust at socket areas.

is recommended that the paint system on the suspender ropes be inspected every 2 years during the biennial inspection.

3.3.2.13 Cable Strands

It is not recommended to paint the cable strands. Oiling of the strands is recommended as an effective means of preventing corrosion of the strands, along with the dehumidification of the anchorages. During the biennial inspection, the strands should be evaluated to determine if the oil is still protecting the strands or the strands need to be reoiled.

3.3.2.14 Eyebar Coatings

A typical eyebar coating regime consists of a three-coat paint system using a 100% solids-penetrating epoxy sealer, followed by two coats of epoxy mastic. A coat of heavy grease is then applied starting at the eyebar–concrete interface and extending 6 in. up the eyebar. In certain cases, the concrete between and adjacent to the eyebars is sealed with two coats of epoxy mastic. It is essential that all rust is removed from the eyebar concrete interface and the eyebar strand shoe.

There are two levels of inspection of the main cable coatings that are recommended:

1. Visual inspection by engineering staff every 6 months: An attempt should be made to visually inspect even the less accessible innermost eyebars, using mirrors or other means, to verify that coating and grease are intact. Any defects in the coating or grease should be noted for immediate correction.

2. Biennial inspections: A visual inspection of the eyebars is made during the biennial inspection to verify that coating and grease are intact. Any defects should be noted for immediate correction.

3.3.2.15 Suspension Tower Coatings

Suspension bridge tower coatings can be subdivided into two categories: exterior coatings and interior coatings. Exterior coatings are typically a three-coat system of zinc-rich primer, epoxy intermediate coat, and urethane/polysiloxane finish coat. The roadway splash zone and areas below the roadway that handle runoff are at the highest risk for coating deterioration, and thus require more frequent inspection and maintenance. Coatings above the splash zone will last for many years with little, if any, maintenance.

Tower interior coatings—the majority of the coatings on the interior of steel bridge towers—are protected from moisture and the harmful effects of ultraviolet light. These coatings should last for many years and are relatively maintenance-free. However, some areas of the tower are prone to ponding, such as the lower struts and the base cells at the tower legs. These areas are coated with up to four coats of immersion-grade epoxy. It is also important to prevent water accumulation by establishing and maintaining weep holes and drainage.

Inspection of the tower coatings should (Figures 3.34 and 3.35) be performed during the biennial inspection.

Figure 3.34 Typical view of tower steel paint that is deteriorated and in need of replacement.

Figure 3.35 Typical view of freshly painted tower.

Inspection of coatings on the exterior of towers is largely visual from the top of towers, main cables, roadway, and tower pier structures. Inspection of the interior tower coatings is usually focused on the areas that are known to have water infiltration issues, such as the base cells, strut areas, and so forth.

ACKNOWLEDGMENTS

The authors acknowledge Joseph Keane, Aris Stathopoulos, and Michael Bronfman of MTA Bridges and Tunnels for their review and comments on this chapter. The authors also acknowledge Dr. Sreenivas Alampalli of New York State Department of Transportation and William Moreau for their review of the chapter and comments.

REFERENCES

1. FHWA. 2004. National Bridge Inspection Standards. 23 CFR Part 650. *Federal Register*, 69(239).
2. Alampalli, S., and Jalinoos, F. 2009. Use of NDT technologies in U.S. bridge inspection practice. *Materials Evaluation*, 67(11), 1236–46.
3. New York State Department of Transportation. 1997. *Bridge Inspection Manual*. Albany: New York State Department of Transportation. With updates through 2006.

4. Alampalli, S. 2014. Designing bridges for inspectability and maintainability. In *Maintenance and Safety of Aging Infrastructure*, ed. D. Frangopol and Y. Tsompanakis. Boca Raton, FL: CRC Press.
5. Federal Highway Administration. 2012. *Bridge Inspector's Reference Manual*. Washington, DC: Federal Highway Administration.
6. National Cooperative Highway Research Program. 2004. *Guidelines for Inspection and Strength Evaluation of Suspension Bridge Parallel Wire Cables*. Report 534. Washington, DC: National Cooperative Highway Research Program.
7. Federal Highway Administration. 2012. *Primer for the Inspection and Strength Evaluation of Suspension Bridge Cables*. Report FHWA-IF-11-045. Washington, DC: Federal Highway Administration.

Chapter 4

Evaluation of Suspension Bridge Main Cables

Barry Colford

CONTENTS

4.1 INTRODUCTION

The main cables of a suspension bridge are the primary load-carrying members and are vital to the structural integrity of the bridge. The main cables transfer the suspended dead load of the deck, and all the transient loads, in tension to the ground via the towers and anchorages. The main cables of the longest-span suspension bridges are usually formed from thousands of very high-tensile steel wires. These wires are produced by cold-drawing steel rods through a series of reducing dies until the final diameter of typically around 5 mm is achieved. The diameter of wire used varies between bridges, but historically 5 mm wire was used for practical reasons, as wire with a diameter larger than this has greater stiffness and requires higher tension for handling during construction. The wires are usually galvanized to provide the first stage of a protective system.

There are two processes commonly used to erect parallel wire cables. In one arrangement loops of individual wire are drawn out across the spans to form the cables. In the other, complete preformed factory-produced parallel wire strands are transported to site on large steel reels, and whole strands are hauled across the spans supported by rollers. Both are mechanical processes that involve handling and transporting the wire to the site and erecting it in exposed conditions using temporary catwalks for access. Both methods inevitably cause damage to the protective galvanizing.

4.2 PARALLEL WIRE CABLE CONSTRUCTION

Aerial spinning is the oldest and most widely adopted method of erecting cables. The process involves hauling loops of wire across the bridge using spinning wheels. The traditional spinning process involved geometric wire-by-wire adjustment from a temporary catwalk or walkway, as shown in Figures 4.1 and 4.2. The controlled tension method more recently developed eliminates the need for wire-by-wire adjustments and is less labor-intensive. The use of an aerial spinning wheel to place wires was invented by French engineers Charles Bender and Louis Vicat in 1820 and improved to achieve greater efficiency by John Roebling, an American engineer of German origin, who used the technique combined with the use of steel wire to form the main cables of the Brooklyn Bridge, completed in 1883. The Brooklyn Bridge, with a span of about 468 m, was by far the longest span in the world at that time and arguably the world's first long-span suspension bridge. This proved to be the catalyst for other engineers designing

Figure 4.1 Erecting a catwalk on Forth Road Bridge.

Figure 4.2 Cable spinning on Forth Road Bridge.

long-span suspension bridges to utilize the aerial spinning method and high-strength steel wire to achieve even greater spans. Aerial spinning is still used today, although significant improvements have been made over the years to try to reduce the time and resources required to spin cables, and thus reduce the cost of construction.

As suspension bridge spans have increased, the need to optimize the operation of spinning the cables, which is expensive in terms of both resources and time on site, became the focus of contractors and engineers. As a result, the use of prefabricated parallel wire strands (PPWSs) has been developed to try to reduce site costs. PPWS is prefabricated in the factory by assembling wires in a hexagonal shape of commonly 61, 91, or 127 wires. The wires are bound with polyester tape at intervals along the length of the strand and socketed at both ends.

Cables constructed using PPWS have no splice ferrules and less crossing wires. Therefore, compaction is improved. The strand is formed in the factory in better conditions than on site, and the erection of the cable is less prone to disruption due to wind. However, very large reels are required to accommodate the strand lengths, and their weight and size make them difficult to transport and handle.

The westbound Chesapeake Bay Bridge (officially the William Preston Lane Jr. Memorial Bridge) in the United States opened in 1973 and is one of the earliest uses of PPWS, although the main span is only 488 m. In Japan, Korea, and China, PPWS has become commonplace on long-span suspension bridge cables constructed in recent years. The main cables on the

Figure 4.3 PPWS cable prior to wrapping and compacting on Akashi Kaikyo Bridge. (Courtesy of Honshu-Shikoku Bridge Expressway Co., Ltd.)

Akashi Kaikyo Bridge, which, with a main span of 1991 m has the world's longest span, were constructed using PPWS. On the Akashi Kaikyo, shown in Figure 4.3, the length of strand was over 4000 m, and the weight around of each large-diameter reel was around 95 tons.

The 1385 m span Jiangyin Yangtze River Bridge in China opened in 1999 and has main cables formed using PPWSs. In the main span, the main cable consists of 169 wire strands, and each wire strand consists of 127 wires with a diameter of 5.35 mm, a strength of 1600 MPa, and a length of 2180 m. The 1490 m span Runyang Yangtze River Bridge, also in China, opened in 2005, and each main cable consists of 184 prefabricated parallel wire strands with a length of 2600 m. Each strand consists of 127 wires with diameter of 5.35 mm and strength of 1670 MPa.

The standard diameter of wire used in nearly all the American long-span suspension bridges built in the twentieth century using parallel wires was nominally 0.192 in. (4.88 mm) ungalvanized. This size of wire was also adopted in the UK on the Forth Road Bridge, Severn Bridge, and Humber Bridge. Table 4.1 provides some values of wire and cable parameters on various bridges formed using parallel wires.

In an aerially spun cable, the wires are formed into strands typically of around 300 to 500 wires per strand, and the strands are then compacted into a circular cross section, ready to accept cable bands and wrapping, using a compacting machine, as shown in Figure 4.4. The theoretical minimum percentage of voids in a perfectly compacted cable is 9.3%, but in practice, this figure is usually more than double the minimum theoretical value, and the exact percentage is determined by measuring the cable

Table 4.1 Wire and Cable Parameters for Suspension Bridges

Bridge	Span, m	Year Opened	Uncoated Diameter, mm (in.)	Specified Wire Strength, MPa (ksi)	No. Wires per Cable	No. of Cables	Comment
Brooklyn	486	1883	4.6 (0.181)	1,096 (159)	5,358	4	Supplementary stays used Rail and road traffic
George Washington	1,067	1931	4.88 (0.192)	1,517 (220)	26,474	4	Road traffic only—two levels
Verrazano	1,298	1964	4.88 (0.192)	1,517 (220)	26,108	4	Road traffic only—two levels
Forth Road Bridge	1,006	1964	4.88 (0.192)	1,544	11,618	2	Road traffic only
Humber Bridge	1,410	1981	4.88 (0.192)	1,540	14,948	2	Road traffic only
Storebaelt East Bridge	1,624	1998	5.38	1,570	18,648	2	Road and rail traffic
Akashi Kaikyo	1,991	1998	5.23	1,800	36,830	2	Road only (PPWSs used)
Runyang Yangtze River Bridge	1,490	2005	5.35	1,670	23,368	2	Road traffic (PPWSs used)

Figure 4.4 Compacting machine.

diameter. The percentage voids within cables will vary between bridges but are usually within the range of 18%–22%.

Following compaction, the cable is usually covered with a protective lead or zinc paste, and then wrapped circumferentially with a mild steel galvanized wire applied under tension using a wrapping machine, as shown in Figure 4.5. Traditionally, this operation is delayed until most of the dead load is on the bridge, as the Poisson's effect of increasing tension in the cable would loosen the wrapping wire if it were applied earlier. The cable wrapping and tightening of the cable band bolts improves the void ratio within the cable considerably.

Figure 4.5 Wrapping machine.

As can be seen from Table 4.1, the ultimate tensile strength of steel used to produce bridge wire has increased significantly since John Roebling specified the bridge wire on Brooklyn Bridge as having a tensile strength of 1096 MPa (159 ksi). The wire used on the Akashi Kaikyo Bridge has a tensile strength of 1800 MPa, motivated by a preference to construct the 1991 m (6532 ft) span using large-diameter single cables, and hence avoid the complications associated with double suspension cables. The increased tensile strength of 1800 MPa was achieved by increasing silicon rather than carbon content, which would reduce toughness. This increase in silicon made steel production more time-consuming and costly, but in addition to increasing wire strength, it has the beneficial effect of halving the reduction in wire strength that occurs during hot dip galvanizing. Wire manufacturers are pushing to produce wire of even higher grades to meet the demand to design longer spans.

The steel used to make bridge wire does have a high carbon content (0.8%–0.85%) in comparison to weldable structural steels. Unfortunately, one of the disadvantages of achieving higher-tensile strengths is the loss of wire ductility—most notably torsional ductility and elongation at failure. High strength and low ductility means that bridge wire is more prone to embrittlement and cracking than would be the case if it were of lower strength and had greater ductile properties.

Most bridge wire used in main cable construction is galvanized to provide corrosion protection. There are some notable exceptions to this, such as the Williamsburg Bridge in New York, where ungalvanized wire was used to form the cables. Typically, galvanizing is applied at 300 g/m, or approximately 40 microns thick, and is applied immediately following cleaning of the wire in an acid bath.

4.2.1 Helical Strands

For shorter-span suspension bridges, it is feasible and more economical to use cables made up of helical strands. Helical strands are manufactured in the shop or factory by helically winding individual wires around a straight inner wire core. Spiral strands are typically used, but in some cases, locked-coil strands are used, with trapezoidal-shaped wires wound round the inner wire core and an outer layer (or layers) formed using interlocking Z-shaped wires.

Because of the twist of the wires in a helical strand, there is a reduction in the modulus of elasticity when compared to parallel wires, and this reduction can be up to 25%. In addition, the ultimate tensile strength of the strand can be reduced by as much as 10%, compared to parallel wires. These reduced parameters are less critical for shorter spans but make helical strands unsuitable for longer spans.

Due to the nature of their construction, helical strands will exhibit non-elastic stretching when first loaded. Therefore, after the strand is socketed, cyclic prestretching has to be carried by applying an axial load exceeding the working load to the strand. However, some settling and bedding in of the strands is perhaps inevitable for this type of rope. This occurred on the Lillebaelt Bridge in Denmark, causing a permanent sag in the deck.

Table 4.2 shows some bridges with helical strand cables.

Various types of arrangements have been used to form the strands into a cable. The most commonly used method is to lay the strands in parallel alongside each other and enclose the bundle by wrapping the cable. Smaller filler strands can be used to aid wrapping, as can filler pieces made from aluminum, plastic, or, in the older U.S. bridges, cedar wood. In the case of the Askoy Bridge, there is an open rectangular array of strands, clamped at the hanger locations.

4.2.2 Recognition of the Problem of Cable Deterioration

The United States has the largest stock of older long suspension bridges, and the problem of corrosion within main cables has been recognized for some time. As early as 1968, the operators of the Golden Gate Bridge carried out a limited inspection of the main cables of the bridge, which indicated that some wires in the outer layers had suffered corrosion. In 1969, the New York State Bridge Authority began unwrapping and evaluating main suspension bridge cables on a 5-year cycle. Broken wires found in the 1980s prompted rehabilitation efforts on the main cables of the Bear Mountain and Mid-Hudson Bridges.

Inspections had also been carried out on a number of bridges over the Ohio River, including the Portsmouth Bridge, Maysville Bridge, and Covington Bridge, in the late 1970s by the Kentucky Department of Transportation. This work followed the closure of the U.S. Grant Bridge over the Ohio River, by the Ohio Department of Transportation in 1978, after severe corrosion was detected in the main cables, after unwrapping, in 1975.

In December 1997, the eminent engineer Jackson Durkee wrote to the Transportation Research Board (TRB) of the National Academies in the United States, expressing concerns over the lack of procedures and guidelines for the inspection and strength appraisal of suspension bridge main cables and pressed TRB to launch a project to address these issues.

In February 1998, Columbia University (Raimondo Betti and Maciej P. Bieniek) submitted a report on the condition of the main cables of 10 suspension bridges on behalf of four bridge authorities in New York. This report summarized the work carried out on all the bridges by others. These bridges were

Brooklyn
Williamsburg
Manhattan

Table 4.2 Details of Bridges with Preformed Factory Strand (PFS)

Bridge	Span	Completed	Type of PFS	No. of Strands	Major Strand Diameter, mm	Comments
Waldo–Hancock	244	1931	Spiral strand	37	35	Galvanized now, demolished due to cable corrosion
I-74 original bridge	226	1935	Spiral strand	37	31 @ 38.1 and 6 @ 25.4	
Chesapeake Bay	488	1952	Spiral strand	61	42	Original bridge eastbound
Rodenkirchen	378	1954	Locked coil	37	69	New cable details
I-74 newer bridge	226	1959	Spiral strand	37	31 @ 38.1 and 6 @ 25.4	
Tancarville	608	1959	Spiral strand	56	72	Old cable details
Tamar	335	1961	Locked coil	31	60	
Lillebaelt	600	1969	Spiral strand	61	69	
Askoy	850	1992	Locked coil	21	99	

Note: Rodenkirchen was widened and a third cable was added, making it unique among suspension bridges in the world. Work was completed in 1994–1995.

On Tancerville's, each main cable was originally made up of 56 individual strands of 169 wires in the main span. There were an additional four strands in the side spans anchored in the saddles at the top of the main towers at the towers and anchorages. Breakages within the 4.7 mm ungalvanized wires were first discovered in 1965, only 6 years after opening. Both main cables were replaced in 1999. The new cables are made up of 90 smaller individual spiral strands using galvanized wire.

Waldo–Hancock Bridge had to be demolished in 2011 due to cable corrosion and was replaced by a cable-stayed bridge.

Bear Mountain
Mid-Hudson
George Washington
Triborough
Bronx-Whitestone
Throgs Neck
Verrazano-Narrows

The study found that despite the age of the bridges, the condition of the main cables was generally satisfactory. Wire deterioration manifested itself as occasional broken wires, wire embrittlement, and usually a very small loss of tensile strength. The factors controlling this deterioration were found to be the quality of the original construction and the quality of maintenance. The loss of cable strength varied from negligible to about 35% (in the case of the Williamsburg Bridge). In no case was the loss of cable strength found to be causing a reduction of the cable safety factor below an acceptable level. In fact, the main concern was the presence of severe corrosion damage to strands in the anchorage and eyebar corrosion.

Perhaps the most important recommendation from this study concerned the maintenance and rehabilitation of the main cables. Specifically, these were as follows:

- Cable unwrapping and opening should be considered at 30 years of service and at a maximum interval of 10 years thereafter. Should significant wire deterioration be detected, in-depth cable inspection should be performed every 5 years.
- The cable-painting interval should be 10–12 years, with at least spot painting as soon as deterioration or cracks in the paint are observed. Repair or replacement of cable band caulking should be included in the painting work.
- Cable rehabilitation should be undertaken when corrosion of wire zinc coating becomes evident, and definitely not later than when the first signs of steel wire corrosion are observed. Cable rehabilitation work should include splicing of broken wires, application of a corrosion inhibitor, installation of new wrapping over red lead paste, tightening of cable band bolts, replacement of caulking around cable bands, and painting.
- Cable band bolt tension should be monitored, and the bolts should be tightened as needed. Loose bolts allow the bands to slip, which in turn changes the bridge's static force condition. Equally important is the fact that even very small movement of the bands damages the wrapping and opens the cable to water penetration. The recommended interval for bolt tension verification and tightening was given as 25 years.

In addition, and importantly, further research recommendations included the development of a system for the evaluation, rehabilitation, and maintenance of the main cables of suspension bridges.

It was not only in the United States that concerns were being raised over the potential for the deterioration of main cables on suspension bridges. In Japan, visual inspections were first carried out on the Honshu-Shikoku suspension bridges between 1989 and 1992. The wrapping wire was removed during these inspections, and corrosion was found in the top layers of wires. No further intrusive internal inspection of the cable appears to have been carried out on these bridges; however, a significant response to these findings was initiated by the Honshu-Shikoku Bridge Authority. This includes a research program to determine the causes of the corrosion and the development of a method to improve the protection of main cables from deterioration.

Following Jackson Durkee's letter and the work carried out on the New York bridges, a National Cooperative Highway Research Program (NCHRP)–sponsored Workshop on Safety Appraisal of Suspension Bridge Main Cables was held in Newark, New Jersey, in November 1998. It was recognized that in the United States, there were almost 50 major suspension bridges, and the majority of them were, at that time, more than 50 years old. It was also recognized that these bridges were critically important to the economic well-being of the country. Given the increasing age of these bridges and the increasing traffic load being applied, the need to accurately determine the remaining service life of the main cables of these bridges, and take any necessary steps to extend service life, was considered to be a priority. There was another important reason to carry out this project, and that was to try to minimize unnecessary failures or unnecessary replacements of main cables due to unreliable methods of inspection and strength evaluation. The highest-priority needs identified were

- The development of cable inspection, sampling, and testing guidelines
- The development of models to predict the strength of deteriorated cables

Weidlinger Associates, Inc. of New York was selected to perform the research and was assisted by the Altran Corporation and Foster-Miller, Inc.

4.3 METHODS OF EVALUATING CABLE STRENGTH AND SERVICE LIFE

A further workshop was held in Irvine, California, in 2002, and owners and operators, consulting engineers, and academics from all over the world were invited to participate. In November 2004, NCHRP Report 534, *Guidelines for Inspection and Strength Evaluation of Suspension Bridge*

Parallel Wire Cables, was published [1]. It is important to note that the investigation and research work carried out for NCHRP Report 534 was specific to main cables of suspension bridges formed from parallel wires.

Since publication of NCHRP Report 534, another method for evaluating the remaining service strength and service life of main cables—the BTC method—was developed by Bridge Technology Consulting and was published in September 2011 by the New York State Department of Transportation and the New York State Bridge Authority, in cooperation with the Federal Highway Administration of the U.S. Department of Transportation (USDOT) [2]. The BTC is a proprietary method developed for use by Bridge Technology Consulting, whereas NCHRP Report 534 is a generic method available for general use.

In May 2012, the U.S. Department of Transportation, Federal Highway Administration, published *Primer for the Inspection and Strength Evaluation of Suspension Bridge Cables* [3]. This document was intended to supplement NCHRP Report 534 and provide a method of standardization for the evaluation of a cable that has been in service for some time. It also provides standard forms to enable engineers to carry out cable investigations. The primer [3] also references the BTC method of evaluation, but does not compare the two methods of strength evaluation.

Therefore, there are now two published methods of evaluating the strength of the main cables of suspension bridges in service—one generic and one proprietary. Other papers have been published on the evaluation of cable strength [4, 5], but these have used the results from one of the published methods and applied limit state or load factor techniques to the results to evaluate safe load-carrying capacity.

Compared to evaluating the strength of an existing suspension bridge main cable, designing a new one is a lot simpler. The new cable is designed by determining the specified minimum ultimate tensile strength of the wire used and multiplying it by the total area, which is the product of the number of wires and the wire cross section. However, the strength of a bundle of wires, even parallel wires, is not, as it appears, the sum of the whole. Tests carried out at the time of construction of the Bear Mountain and Ben Franklin Bridges [6] on bundles of parallel wires between 37 and 306 individual wires showed that the efficiency of a wire bundle, that is, the tested strength of the bundle divided by the mean tensile strength of the individual wires multiplied by the total area of the strand, varied between 94% and 97%. Therefore, a bundle of individual parallel wires does not have quite the same strength as the sum of the individual wires.

Other tests have shown that when a strand of parallel wires is taken to near its ultimate capacity, individual wires start to break sequentially, and the maximum load is only reached after some of the wires have broken. This fact has to be considered when considering cable strengths.

NCHRP Report 534 [1] provides detailed guidelines, based on previous experience from opening up a number of bridge cables, on all the stages of the work necessary to open up, inspect, and evaluate the strength of a cable. The procedure set out in the BTC method is less prescriptive, as the philosophy underpinning the inspection and testing is based on eliminating possible sampling bias caused by visual assessment, and by using random inspection and sampling to reduce or eliminate that bias.

4.4 CAUSES OF DETERIORATION AND STRENGTH LOSS WITHIN MAIN CABLES

The causes of the deterioration of high-tensile galvanized wires forming the main cables of suspension bridges have long been the subject of study and debate. During work carried out by Hopwood on the Ohio River Bridges in the United States [7], the following four stages of deterioration in galvanized bridge strands were established. These categories were developed for wire strands, but can equally apply to individual wires.

Stage 1: The strand is in "as new" condition. The zinc coating has a bright metallic appearance, though some slight spot corrosion of the zinc may be evident in the form of a thin, white, powdery coating.

Stage 2: The strand is in good condition, and exposure to the atmosphere has given the zinc a dull-grey appearance. The white corroded-zinc film may be present near the interfaces and on the exterior surfaces of the wires. If the white film is removed by scraping, no rust is evident. The second layer of wires may be in worse superficial condition than the outer layer, but as long as the outer layer of wires is stable, the interior wires will probably remain structurally sound.

Stage 3: Much of the strand is covered with a thick, white zinc corrosion product. Spotted rust is also visible on the wires. When the corrosion product is scraped off, the steel under the surface reveals some rust and pitting. Wire breakage is possible during this stage; however, the breaks will not be clustered in large numbers, except near points of high-stress concentration.

Stage 4: The strand will be severely rusted and pitted. Some zinc corrosion will be displaced by corrosion of the underlying steel (rust). The wires will have a speckled brownish red and white appearance. If loading and corrosion conditions are severe, the strand will develop many fractures and will eventually fail.

Testing carried out from a study of two bridges in the United States [8] on wires exhibiting all four stages of deterioration showed three types of wire degradation:

Type 1: Consists of shallow pitting from the underlying steel after zinc depletion

Type 2: Shows similar shallow pits as type 1, but with well-defined cracks within the wire cross section

Type 3: Consists of well-defined cracks through the cross section of the wire, but without any surface pitting and with the zinc coat intact

This study showed that significant cracking can occur in bridge wires where there are only localized nonpitted defects within the zinc coating and where the adjacent coating remains intact. From the numerous tests that have now been carried out on a number of wires sampled from bridges throughout the United States, Europe, and Japan, the following observations have been made:

- It is the presence of cracks within wires that causes the significant proportion of strength loss within a cable formed from parallel galvanized wires.
- These cracks tend not to become visible until the wire breaks.
- Wire cracks and breaks can occur well before any significant loss of section to corrosion.
- Corrosion on its own is not the cause of wires cracking, and hence breaking, leading to strength loss.
- Tensile testing of galvanized bridge wires that exhibit significant corrosion has provided results that show that the wires have not suffered any loss of tensile strength or loss of ductility, and wires that are less corroded have, when tested, been significantly less ductile. In the failures that have been observed in cables with galvanized wires, it would seem that corrosion could be regarded more as an indicator of the presence of moisture within a cable, rather than the reason for loss of strength. The presence of moisture is required to produce the atomic hydrogen, which appears to be the key to cracks occurring and enlarging to become critical.
- Obviously, if corrosion was allowed to continue within a cable even without the formation of cracks, loss of section would eventually occur, resulting in wire failure.

4.5 INVESTIGATION OF THE CONDITION AND APPRAISAL OF STRENGTH OF MAIN CABLES

Regular, systematic, and thorough inspection of bridges and other highway structures is essential in order to highlight any defects and to ensure that any maintenance or remedial works are minimized and do not affect the structural integrity of the bridge or highway structure.

For simple noncomplex bridges, inspections can be carried out on a regular frequency for all components. This can be done by carrying out a general inspection every 1 or 2 years and a more detailed inspection every 5 or 6 years. Special inspections would be carried out after significant events or incidents that affected the structure, such as extreme winds, flood, or accidents. However, this approach is not suitable for more complex structures, such as suspension bridges, and many owners and engineers are now adopting an inspection regime that identifies the criticality and vulnerability of each component of the bridge and allocates a score to each of these parameters. A matrix is set up with numerical values assigned to each individual element. The element is then inspected according to the product of the criticality and vulnerability score. The more critical and or vulnerable elements produce a higher score, and the inspection frequency is higher. If there are regulations, standards, or guidance covering a certain element that determines inspection frequencies, then that is taken as the minimum.

NCHRP Report 534 [1] sets out inspection frequencies in the opening chapters based on federally required biennial hands-on inspection of non-redundant elements [9, 10]. However, it is imperative that the main cables on all suspension bridges are subject to a rigorous internal inspection after a set number of years in service, and thereafter at a frequency determined by the condition found and any intervention carried out. The federal and state inspection guidelines also realize this and require specialized procedures' inspection of complex bridges [9–11]. This crucially important message is set out in the NCHRP report, and guidance on inspection and inspection frequencies is given.

Perhaps, the most fundamental and important advice given within NCHRP Report 534 [1] is

- Internal inspections of main cables are necessary at some point in the life of the bridge
- Regardless of the external condition of the main cable, the first inspection should be carried out no later than 30 years after opening

This timing of the first inspection will be increasingly open to interpretation, given that the installation of cable dehumidification systems on new suspension bridge cables has become standard practice in recent years. The cables will have been compacted, wrapped, and (hopefully) suitably protected prior to the opening date of the bridge. However, the presence of external signs of corrosion, dissimilar metals, or other factors might mean that the first internal inspection is carried out before 30 years from opening date. This will be for the owner and engineer to determine based on the particular facts and using engineering judgment.

The main cables of a suspension bridge, being made up of very high-tensile wires, have a much higher capacity to support loads than other structural steel members. However, suspension bridges, by the very nature of their structural form, do not have a high degree of load redundancy, and this especially applies with regard to the main cables and anchorages. If a main cable or cable anchorage were to fail, then catastrophic collapse of the bridge would follow.

This was, perhaps, the reason that Roebling and those other early bridge engineers designed main cables with a factor of safety (FOS) against ultimate failure greater than 4. Given the deterioration that has been found in some cables, the decision to use a high-FOS figure has proven to be extremely prescient. When the construction of long-span suspension bridges started in Europe in the 1960s, the factors of safety adopted dropped to as low 2.2, because of economic pressures and as engineers concluded that their methods of analysis and determining loads were more advanced than those of their predecessors. Given the deterioration that has been found in some cables from bridges built in the 1960s, the adoption of such low factors of safety has, with hindsight, proven to be an unwise decision.

The issue for suspension bridge owners and engineers is how to inspect the thousands of individual wires over their full length within the main cables, when they are hidden from view by wrapping wire, paint, or adjacent wires. The main cables and their anchorages are the most critical part of a suspension bridge, and it has been shown that an external examination of the cables, no matter how detailed, is not good enough to determine the condition of the cable or give an indication of the residual life of the cables. What is now also evident, from the work carried out at Forth Road Bridge and other bridges, is that even if a main cable does not exhibit any signs of deterioration on the outside, this does not mean that significant deterioration is not taking place underneath the paint and wrapping wire.

It is, of course, not possible to examine every part of the surface area of each individual wire. Therefore, no matter how detailed the inspection is, some extrapolation of the findings will be required to determine the likely condition of the whole cable. The whole purpose of the inspection is to determine current condition and whether or not any reduction in original strength has been due to deterioration. Therefore, an internal inspection is vital.

The results of an internal inspection can be used to determine the current capacity of the cables and their estimated residual strength and service life. It is not possible to determine the actual properties of each individual wire in the cable. Indeed, the condition, and therefore the mechanical strength of each wire, may vary along the length of the cable. As a consequence of this, the evaluation of current and future strength will be a probabilistic calculation with a relatively significant error because of the small sample size.

There have been many proposals put forward to inspect the main cables of suspension bridges using nondestructive testing and remote monitoring techniques, including radiography and techniques adopting electromagnetism in some form or other and acoustic monitoring. With the exception of acoustic monitoring, none of these techniques have proven to be effective on the relatively large-diameter cables found on suspension bridges. Electromagnetic techniques have had some success on smaller-diameter strands and ropes, but only up to 100 mm in diameter. The technique has been used with some success along the free length of suspenders. However, it is less effective at the sockets, and a further drawback is that the technique can only determine the potential loss of cross section and cannot detect wire cracking or breaks.

One of the earliest recorded uses of acoustic monitoring was on Tancarville Bridge in France. Sonic detectors were installed after broken wires were detected in the cables. Acoustic monitoring has now been adopted on a number of bridges in recent years with some success. However, this form of monitoring will only provide information on wire breaks as they occur; it cannot provide information on historical or future wire breaks. In addition, there does appear to be the need for further validation of the efficacy of acoustic monitoring on main cables. The usual tests to calibrate and validate an acoustic monitoring system are by hammer testing and by the actual cutting of a wire. However, the energy released by a broken wire will be somewhat different than that of a cut wire, and perhaps some deteriorated wires break with more energy than others. There is some limited anecdotal evidence of actual wire breaks being noted by engineers during cable unwrapping and these being picked up by the acoustic monitoring system. Further research is also required to determine the optimum spacing of sensors on the cable.

Acoustic monitoring, like any monitoring system, should not be used as an alternative to inspection, but it can be a useful warning system and provide guidance on where to inspect on the cable. It must be emphasized that the inspection regime of any structure, especially cable-supported structures, must form part of a formal and well-defined maintenance regime. It is recommended that these large cable-supported bridges have their own maintenance or engineering manual.

The first investigation into the condition and strength of the cables of a main suspension bridge can be divided up into a number of stages:

- Desk study
- External inspection during cable walk
- Internal inspection and sampling
- Laboratory testing
- Evaluation of current and future strength of the cable

There are two important points to note. First, it is vital to appoint an engineer and a contractor who are both competent and experienced in this type of work. Second, it should be accepted that any internal inspection disturbs the cable, and it is not possible to reinstate exactly back to the original condition. In addition, more ferrules from sampling wires will be inserted in the cable, causing further voids. Therefore, it is imperative that the important work of recompacting and rewrapping is carried out with great care and attention, and sufficient supervision by the engineer is afforded to that task.

It is preferable if cable investigations are not carried out as "least-cost bids," and that some form of meaningful quality assessment is adopted within the tendering process.

A desk study to determine as much information as possible about the cables should be carried out by the engineer as a first important task. The study should try to include the following:

- Details of the construction of the cable. Mill certification of wire strengths and properties is not always available but is very useful. Where the wire was stored, the conditions during storage, and the size of reels used to transport wire are all useful intelligence.
- Details of any issues during construction, such as high winds causing wire entanglement, rejection of material, labor disputes that might have caused delays, and any particular issues at the saddles or anchorages. The compaction records are also important to determine what void ratio was thought to have been achieved.
- The inspection and maintenance records since opening, including painting records; details of any water leaching from the cables; any weathering, damage, or degradation of the coating system; any corrosion apparent on the wrapping wire; and any blocked drainage holes in the cable bands. The cable bands, tower top saddles, and anchorages are points of obvious weakness, where the cable has to be caulked as it enters a housing, as all caulking materials have a limited life. Any history of leakage in these areas should be noted.
- The dead load should be recalculated to ensure that it is as accurate as possible and takes into account any additional load that has been added since opening. The determination of the traffic load will require some thought. There will be the original design live load and the national standard live load, which would be applicable to new construction. There could also be a national standard for assessing bridges, which is likely slightly less than the current design standard, and a bridge specific loading obtained using a weigh-in-motion system. Discussion between the owner and the engineer will be required before determining which live load to use. It may be necessary to construct a new structural model of the bridge in order to determine the most accurate and up-to-date suspender and cable loads.
- The historical and current level of air pollution adjacent to the bridge site is a useful indicator in determining the likely level of deterioration of the wires.

The desk study should normally be concluded prior to carrying out the next stage of the investigation, which is to carry out a cable walk to confirm some of the desk study information, and also to look for other obvious external defects, such as rust staining, dripping water, loose wrapping, or surface ridges indicating crossed wires.

Once the desk study and cable walk have been completed, a number of panels (a panel is the length of the cable between the suspenders) have to be selected for inspection. Those panels where external inspection has indicated possible internal deterioration should be selected for unwrapping, wedging, and inspection. These are likely to be the panels that are most at risk. If there are no external signs of deterioration or possible causes of deterioration, the recommendations set out in NCHRP Report 534 for a first internal inspection are that the cable should be opened and inspected at a minimum of three locations along each cable [1]. The BTC method recommends that for a cable where there is no previous history of inspections and no evident signs of deterioration, panels should be randomly selected [2].

The investigation should also include an inspection of the anchorages. Cable sleeves adjacent to the towers or cable bents should be removed and the wires examined.

The primary method of carrying out an internal inspection is to remove the wrapping wire and open up the cable by driving hardwood or plastic wedges between the wires, and this technique has been used for over 40 years in the United States by various owners and engineers. Unfortunately, opening up the cables of a suspension bridge is not an easy task, and thus is expensive. It can also involve disruption to users of the bridge if lane or carriageway closures are required.

The arrangements for carrying out the work, whether it is done in-house or by a contractor, will be a decision for each owner to make and will depend on the owner's resources. One of the most important decisions is the type and form of access and containment that is to be provided. This will be influenced by many factors, including the existing access arrangements on the cable, the layout of the deck and proximity of traffic, and most importantly, the owner's "risk appetite." There are numerous health and safety requirements to be met that are likely to include dealing with the containment and disposal of old red lead paste, which may be brittle and friable.

One of the great challenges of working on suspension bridges is that the main structural elements, the cables, are directly above the heads of the users. It is not possible to eliminate risk to users when working on the main cables if traffic is kept running. Unless the bridge is completely closed to pedestrians and all traffic, any work on the cables, whether it is only inspection or minor maintenance, immediately involves increasing the risk to the users below. The risk can be significantly reduced if only the carriageway below the main cable being accessed is closed. However, there is still a risk of something falling and bouncing over to the live carriageway, and on each bridge that risk will be different, depending on the layout of the deck. What society expects is that the owner and engineers will take reasonable steps to minimize that risk. Of course, the ultimate arbiter of what is reasonable is likely to be the judicial system, and so when choosing, or accepting, a proposed access system, it is recommended that a full risk assessment be carried out to identify all risks and take all reasonable measures to mitigate against the most significant risks.

The other important issue to keep in mind is that although the project is primarily an investigation, a significant amount of plant and machinery will be used in the works. The access platform will have to be capable of supporting a large compacting machine and a wrapping machine, and both machines will have to be loaded onto and taken off the platform.

The panels to be opened, for a first investigation, are likely to be spread across all the bridge spans and on both cables. Access and containment at the low points of the cable are usually straightforward, as scaffold and containment sheeting can be readily erected. Therefore, it is advisable that the first panel opened is at a low point in the cable to allow the procedures and processes to be trialed in an area with easier access and containment. Discrete access is required at each panel point using scaffolding at the deck level or fabricated steel platforms higher than the deck. Access to the platforms at a high level can be slow (either cradle access from the deck or cable walking). However, it can provide the most stable platform and the best containment envelope. The platforms can be winched onto the cable using overnight closures, causing minimum disruption to traffic. The wrapper and compactor can be lifted on in one piece at the deck level and winched up with the platform. Steel is the preferred material, and as a result, the platforms can weigh in excess of 12 tons on an 18.29 m (60 ft) panel. The number of platforms manufactured will depend on the contractor's schedule and resources. This type of access was used successfully on Forth, Severn, and Humber Bridges in the UK (Figure 4.6).

Figure 4.6 Discrete cable access platform.

Other forms of main cable access used for cable inspections carried out as part of more significant works, such as rewrapping, painting, or installing cable dehumidification operations, are as follows:

- Provide a walkway/catwalk over the full length of both cables: This option has the best flexibility if the investigation needs to be extended, and if other works on the cable are required. However, such systems were originally intended to provide the access necessary to spin a new cable without the inconvenience of traffic below. Therefore, there is a reduced containment provision. In addition, the robustness of the system in high winds would also have to be considered. This method is illustrated in Figure 4.7.
- Provide a cable crawling platform or vehicle that is pulled or powered along the cable or hand strands: This would provide the necessary flexibility in the event that further panels need to be opened. However, the platform needs to be split in the middle to run past the suspenders, and this causes an issue with space to accommodate the wrapper and compactor. For this reason, this option was not adopted on the three UK long-span suspension bridges, but was used successfully to install the dehumidification system on all three bridges (Figure 4.8).

The actual physical work on site to facilitate the inspection involves gaining access to each panel for a sufficient period to allow the following operations necessary to complete the work:

- Installing the access system
- Unwrapping the cable

Figure 4.7 Temporary walkway.

Figure 4.8 Movable cable access platform.

- Removing any old red lead or other protective paste and cleaning the cable
- Measuring the diameter of the cable
- Wedging open the cable at a number of points around the circumference (Figure 4.9)
- Inspecting the wires within the wedge lines
- Removing samples for testing
- Swaging in wires to replace those broken or removed for testing
- Removing wedges and recompacting the cable, checking the compacted diameter
- Applying new protective paste—ensuring sufficient thickness around the underside of the cable
- Rewrapping the cable, recaulking at the cable bands or flashings as necessary, and applying the protective system
- Removing the access system
- Estimating cable strength and reporting

Prior to the general removal of the wrapping wire within a panel, the original diameter and circumference of the cable should be measured on top of the cable wires. In order to do this, three short lengths (around 50 mm) of wrapping should be removed at approximately 1 m from both cable bands and at the midpoint of the panel (these measurements are then checked during recompaction of the panel after inspection). Any existing red lead paste is removed using plastic mallets. The wrapping wire should not be removed entirely over the length of the panel, as it would be difficult

Figure 4.9 Wedging a cable for inspection.

to recompact the cable close to the cable bands. NCHRP Report 534 [1] recommends leaving the wrapping wire intact over a length of 1.5× the cable diameter, from the cable bands. In practice, it is usual to leave a 1.0 m length of wrapping wire intact at each end. The wrapping wire can be easily soldered to hold the tension at the cut end.

The recommendation set out in NCHRP Report 534 [1] is that during the first inspection, the panels should be unwrapped and opened for a length of at least of 16 ft (4.88 m) and wedged at four equally spaced locations around the perimeter of the cable. If wire corrosion is more severe than stage 2, then the wrapping wire removal should be extended to the point about 1.0 m from the bands to facilitate specimen removal and to allow the wedges to be driven deeper into the cable. The recommendation is that a further four wedge lines should then be driven, and these should also be equally spaced around the cable. If eight wedge lines are required, they should be spaced at every 45° around the cable circumference. Supplementary wedge lines may be driven in areas where significant corrosion is present. The procedure set out in the BTC method [2] is less prescriptive, as the philosophy underpinning the inspection and testing is based on eliminating possible sampling bias caused by visual assessment and by using random inspection and sampling to reduce or eliminate that bias. The eight-wedge pattern is shown, but it is noted that the wedge pattern could be selected randomly.

Once the existing wire wrapping is removed, the main cable can be cleaned and then wedged open for inspection using a bronze starter tool and a combination of hardwood and plastic wedges to hold the grooves open. The wedges can be driven to the center of the cable, but this is not always necessary. The wires on either side of the wedge are exposed for

Figure 4.10 Exposed wires in a cable.

inspection. It should be borne in mind that only a portion of the surface of the exposed wire can be inspected, as shown in Figure 4.10.

It is likely that the bottom of the cable will have suffered the worst deterioration, and the first wedge line should start at the bottom in the middle of the cable and work outward toward the cable bands.

It is possible to drive more than one wedge at a time at different locations around the circumference, and this decision will be dependent on the cable condition, the drivability of the wedges, and the inspection resources. One point to note is that driven wedges can pop out of the cable of their own accord, and it is advisable to secure all wedges with lanyards or straps, and that the containment is sufficient to retain any wedges that work loose.

On completion of driving each line of wedges, a detailed inspection is carried out on the condition of the wires along the cable by categorizing them into the four main stages of corrosion following the descriptions set out by Hopwood [7]. Both NCHRP Report 534 [1] and the BTC method [2] use similar but subtly different descriptions of each of the four stages of corrosion. The NCHRP Report 534 and BTC classifications are shown below.

NCHRP Report 534: Classification of stages of corrosion
    Stage 1: Slight deterioration of zinc galvanizing
    Stage 2: Extensive depletion of zinc galvanizing
    Stage 3: 0%–30% surface ferrous corrosion
    Stage 4: Over 30% surface ferrous corrosion

BTC Method: Classification of stages of corrosion
    Stage 1: The zinc coating of wires is oxidized to form zinc hydroxide, also known as white rust.

Stage 2: The wire surface is completely covered by white rust.
Stage 3: There is an appearance of a small amount (20%–30% of wire surface area) of ferrous corrosion due to broken zinc coating.
Stage 4: The wire surface is completely covered with ferrous corrosion.

Typical stages of corrosion are shown in Figure 4.11.

The inspection procedure set out in the NCHRP Report 534 procedure relies on a visual assessment of the condition of the wires, coupled with laboratory analysis of the tensile strengths of sample wires taken from the cable at various stages of corrosion. This is a crucial part of the NCHRP Report 534 procedure. It relies on engineers visually examining the wires within the wedge lines and determining the stage of corrosion of that wire. Therefore, it is essential that engineers experienced in this type of work carry out the visual inspection. The inspection must be meticulous, and enough time has to be given to allow it to be carried out.

As might be expected, it is relatively easy to determine the difference between a stage 1 wire and a stage 4 wire. However, it can be difficult even for experienced engineers to determine whether or not a wire should be categorized as stage 3 or 4. This can make a significant difference to the estimated strength of the cable. It is an obvious fact that even two engineers experienced in this type of work will examine a cable and may classify the stages differently, especially in stages 3 and 4. This can cause issues if owners require an independent review or verify the investigation. Enlarging the sample size, especially of stage 4 wires, can help reduce the error.

However, when using the NCHRP method [1], the number of samples to be taken for each corrosion stage is specified according to the overall percentage of each corrosion stage found within the worst-affected panel. In addition, there are specific criteria set down for the location of sampling within the cable. In the BTC method [2], the random sampling method is

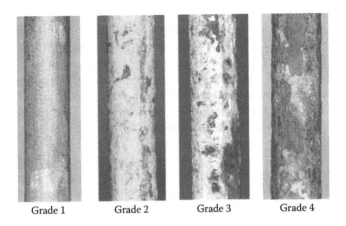

Grade 1   Grade 2   Grade 3   Grade 4

Figure 4.11 Stages of corrosion.

Figure 4.12 Removing a wire for testing.

used so that each wire within the available pool (or accessibility for sampling) has an equal known chance of being sampled. Figure 4.12 show a sample being removed from a cable.

4.6 CABLE STRENGTH EVALUATION

The calculation of the current strength of main cables that have suffered deterioration is complex and challenging. If broken wires are discounted, then it is the presence of cracks within wires, rather than corrosion, that has the biggest effect on strength loss within a cable formed from parallel galvanized wires. Bridge wires that have cracked have a reduced and highly variable tensile strength. They fracture with no reduction in area or necking.

It must also be emphasized that the strength of the cable is evaluated only in the panels inspected. Therefore, if the most deteriorated panel is not evaluated as part of the investigation, then the capacity of the cable is likely to be overestimated.

The mechanical properties of bridge wire are different from normally used structural steels. The stress–strain curves for typical bridge wires are shown in Figure 4.13. As can be seen, there is no distinct yield point. As strain increases, and the curve leaves the elastic range, there is no flat plateau, as would be found in structural steels. Instead, the capacity to take load increases, due to a form of strain hardening, until the tensile strength is reached and the wire fails. The ultimate tensile strength coincides with the point of ultimate strain.

Previous internal inspections have shown that when a wire breaks within a cable, the gap formed between the ends of the broken wire is equal to or slightly greater than what the elastic stretch of the wire would be if it were loaded under dead load. This has led to the assumption that all the wires in the cable are effectively clamped at the cable bands under dead load (which forms the majority of the load in a cable), and thus subject to equal strain.

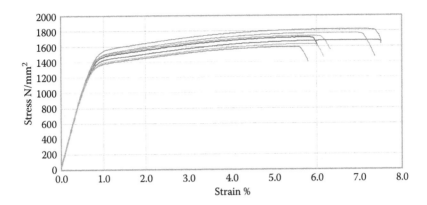

Figure 4.13 Variations in new bridge wire stress–strain curves (due to varying carbon content).

The NCHRP method of evaluating strength is to sample and carry out tensile testing of wires based on predetermined percentages from the different stages of wire corrosion found during the internal inspection. Given the relatively small length of cable that can be inspected, and that samples can only be taken from the wires nearest the surface of the cable, this limitation leads to a significant known error, which must always be remembered when determining the estimated strength.

In NCHRP Report 534 [1], a number of models are considered for evaluating the strength of a main cable, and all are based on the assumption that a wire can carry load up until it reaches the ultimate strain value for that wire. The model used on strength evaluations carried out on the bulk of the bridges has been the brittle wire model. It is assumed when using the brittle wire model that each wire has a limited strain capacity at which it fails and no longer contributes to the capacity of the cable to carry load. The strain capacity of a cracked wire is significantly different than that of an uncracked wire, and hence this model has been questioned by some in the literature.

The description *brittle wire model* is perhaps a bit misleading. It does not mean that the wires lack ductility; rather, it is meant to describe the sudden failure of an individual wire when it reaches its wire-specific ultimate strain.

Consider a cable that has deteriorated and contains corroded, cracked, and broken wires:

- If the cable is subject to a strain just equal to the ultimate strain capacity of the weakest wire (i.e., the wire with the lowest value of ultimate strain), then the force in each wire at that applied strain will vary, and the force in the cable as a whole is the sum of all those individual wire forces.
- The assumption made in the brittle wire model is that each intact wire can resist stress only up to its tensile strength, and that strength

is different for each intact wire. When the stress in the cable reaches the tensile strength of an individual wire, then that wire will break. Once this happens, the wire no longer carries load, and that load is assumed to be distributed equally among the remaining intact wires. Since all the load or stress in the wires is assumed equal at any point, the cable force is taken to be the wire stress multiplied by the area of unbroken wires, and the maximum cable strength is only attained after a number of wires have broken.

- To determine the maximum cable strength, the stress in the cable is assumed to be increased in steps, and at each increment the number of wires that have reached their tensile strength and have failed is determined. The number of newly failed wires at each increment is deducted from the total number of previously intact wires to give the remaining number of intact wires. The total cable strength at the given increment is the area of unbroken wires remaining multiplied by the stress in the wires at that level. As the stress level is increased at some particular increment, the wires will fail faster than the wire force can increase and the maximum cable strength will have been reached.

- Of course, it is not possible to determine the actual strength of a real cable by testing each wire within the cable to determine the individual stress–strain curve, which in any case may vary along the length of the cable for each wire.

- Therefore, a limited number of sample wires are removed from the cable for testing, and tensile strength distribution curves are derived for each of the previously identified corrosion stages (stages 1–4).

In the BTC method [2], random sampling and reliability-based analytical techniques are used to assess the remaining cable strength. The exact method of calculating cable strength is only known to the owner of the patent of the BTC Method. This does make independent verification of strength loss assessment problematic. The probability distribution for wire mechanical properties such as strength are established along with the loads, and then the cable failure mechanism is developed and the serviceability of the cable is determined. After the random sampling of wires from each investigated panel, the samples are mechanically tested to determine the probability of the occurrence of broken and cracked wires. Fracture toughness is used to assess the ultimate strength of the cracked wires and assess the remaining strength of the cable in each panel.

The parameter most bridge owners are interested in is a reliable future estimate of the likely strength of the main cables on their bridges. As the cable deteriorates, more wires become cracked and broken, and this has a significant and increasing effect on cable strength, which makes the strength against time relationship nonlinear.

In the NCHRP guidelines [1], a nonlinear model for extrapolating the cable strength into the future is given, although it is stated that the method is only suitable to predict the strength to only about 10% of the current age of the bridge. Therefore, if the first inspection and strength evaluation are carried out when a bridge is 30 years old, then the future strength would only be projected for a period of just over 3 years. This makes long-term planning quite difficult. It is possible (if desired) to calculate strength degradation further into the future; however, the projected rate of degradation should not be considered definitive. It is prudent to adopt a cautious approach when predicting future rates of deterioration, particularly as the prediction is based on the investigation of a limited number of panels.

The BTC method [2] appears to use a probabilistic-based methodology to assess the remaining service life of the cable. The rate of change of cracked and broken wires detected over a time frame is determined, and the rate of change in effective fracture toughness over the same period is also measured. These rates of change to both parameters are applied to a time-dependent strength degradation prediction model.

4.7 METHODS OF PREVENTING CORROSION

The use of wrapping wire and paste, then overcoating with a paint system, does not keep moisture out of the main cables of suspension bridges. Even on bridges with very good cable maintenance regimes where repainting and recaulking have been carried out regularly throughout the life of the cables, significant deterioration has been found following internal inspection. Main cables move significantly during high winds and seasonal temperature variations, and paints and sealants do not have the elasticity or durability required to accommodate that movement and withstand exposure to the elements. The development and use of S-wire for wrapping increases the chances of excluding moisture, but is not sufficient on its own.

Neoprene, plastic, and other wrapping materials have been used on main cables, often following the discovery of deterioration, to try to keep moisture out of the cables. Unfortunately, experience has shown that this is not very effective and can actually aid the retention of moisture and can also potentially be a barrier to the detection of corrosion.

The corrosion inhibiters, commonly linseed oil or linseed oil-based compounds, have been used by some bridge owners, mainly in the United States, to slow down or halt corrosion within main cables. Oil can be introduced into the cable at the tower saddles, and the expectation is that it will flow down the cables, between the wires, and provide protection against further corrosion. It is not an easy operation to carry out, and the aim is to surround as much of the surface area of the wires as possible with oil. Experience has shown that oil leakage and bulging of the cables can occur at lower levels. There does not seem to have been much recent oiling carried out on main cables, perhaps reflecting the doubt among owners and engineers over the effectiveness of this method. However, for a period of time, oiling was the only viable option available to bridge owners.

That changed following the discovery of corrosion of wires under the wrapping wire on the Innoshima Bridge in Japan, only 6 years after the opening of the bridge. The owners, the Honshu-Shikoku Bridge Authority (HSBA), unwrapped cable panels on the Seto, Ohnarutou, and Ohshima Bridges and found similar corrosion. The HSBA initiated several investigations and accelerated testing carried out on several combinations of wrapping wires and materials, paints, and pastes to determine if it was possible to make a cable completely watertight. The conclusion was that it was not possible to keep moisture out of main cables. The HSBA also concluded that maintaining a dry state within the cable by some artificial means was the only way to protect cable wires against corrosion and investigated various means of achieving this. The work in Japan showed that new galvanized wires would not corrode in an atmosphere where the relative humidity was less than 40%. They examined ways of permanently reducing the relative humidity within main cables, and the artificial means they developed was to inject dry air into and along the length of cables via the interstices between the wires in the cable.

Dry-air or dehumidification systems have now been retrofitted on a large number of bridges in Japan; on the Forth, Severn, and Humber in the UK; on the Hogakusten and Littlebelt Bridges in Scandanavia; and on a number of suspension bridges in China. Systems have now been installed on the Chesapeake Bay Bridge, and the installation of many more systems is planned on bridges in the United States. Systems have been installed on a number of new suspension bridges, such as Hardanger in Norway. Dehumidification system is also being installed on the Third Bosphorus Bridge and Izmit Bridge, both of which are currently under construction in Turkey.

The system works by blowing air, at low pressure, with a relative humidity (RH) less than 40%, into the cable at various inlet points along the length. The dry air travels along the cable and exits at various exhaust points. The cable has to be sufficiently airtight along its length and at the cable bands, tower top, and other saddle covers. The spacing of the exhaust points is crucial for the effective operation of the system, and the experience from other bridges should be sought, which may have to be supplemented by trials. The air pressure used is fairly low, typically about 3000 Pa, as sealing the cable, especially at the cable bands and covers, can be difficult. There are various methods now used to wrap the cable in order to keep the dehumidified air from leaking through the wrapping wire and paint. Elastomeric wrapping material is widely used in Europe, the material being laid around the cable on top of the existing wrapping wire and paint with minimal preparation, as shown in Figure 4.14. In Japan, elastomeric paint has been used in lieu of a wrapping material on some bridges. Given the importance of the protective nature of the wrapping wire and the part it plays in helping maintain the friction between the bridge wires, and the compaction of the cable, it should only be removed locally at the inlet and exhaust ports.

The relative humidity at the exhaust (as well as flow, pressure, and temperature) is measured and monitored; the aim is to achieve less than 40%

Figure 4.14 Applying the elastomeric wrap.

RH at each exhaust. Figure 4.15 shows a typical exhaust. It is assumed that even if all the moist air in the interstices between the wires is not replaced with dry air, in the longer term, due to the formation of a moisture gradient, all the moisture will migrate toward the drier areas and the whole cable will eventually dry out.

Dehumidification clearly does not allow the cable to regain strength lost due to the wire breaks. The expectation is that it will prevent wires at condition stages 1 and 2 deteriorating further and becoming stage 3 or 4 wires, and thus reduce the rate of strength loss and eventually the absolute

Figure 4.15 Typical dehumidification exhaust.

loss of strength. It is difficult to state definitively that dehumidification will prevent future strength loss. From previous research and investigations, it can be seen that cracks occur in wires with very little apparent corrosion or damage to the galvanizing, and it has also been found that wires classified as having stage 4 corrosion can have the same tensile strength and ductility as when they were new. However, perhaps what dehumidification is doing in eliminating moisture is preventing the formation of atomic hydrogen, which may be critical in the formation of cracks leading to wire breaks.

Some bridge owners, who have retrofitted dehumidification systems and have acoustic monitoring systems installed, are reporting that the frequency of wire breaks recorded by the monitoring system has decreased as the cables have dried out. From post-dehumidification inspections carried out on the Forth Road Bridge and the Kita Bisan-Seto Bridge in Japan, it appears that the condition of the cables has not deteriorated since the retrofitting with the dehumidification system. Notwithstanding these results, it is hoped that future inspections on bridges that have had dehumidification systems retrofitted will provide further confidence that dehumidification is effective in halting or slowing down strength loss in cables that have suffered deterioration.

4.8 CONCLUSIONS

The corrosion found within the main cables of suspension bridges has not only caused an immediate problem to the owners and operators of those bridges, but also caused concerns within the engineering community and the wider general public.

Two well-known and well-publicized closures of suspension bridges due to cable deterioration, the Grant Bridge and the Waldo-Hancock Bridge, are salutary lessons for all those involved in long-span bridges. These bridges were both located not in an impoverished third world country, where resources and knowledge may have been scarce, but in the United States. They highlight the absolute need for properly funded and well-organized maintenance of suspension bridges in order to protect these vital national assets, and that regular investigations, including internal inspections of the main cables, must be carried out over the full service life of the bridges. These investigations cost time and resources, but the engineering profession has a duty to ensure they are carried out.

One consequence of the problems that have been found within main cables has been a seeming reluctance to build new suspension bridges. Cable-stayed bridges are now being built with spans exceeding 1100 m. However, at these spans, tower heights are significant, as are compression loads that have to be resisted by the deck. Two main advantages put forward for the choice of cable-stayed structures are that they are cheaper, and that as the cable stays are individual elements, strands or stays can be replaced separately without disrupting traffic. However, given the role of the deck as a global compression member in a cable-stayed bridge, one reason for the less enthusiastic use of cable-stayed bridges in the United States appears to the experience there of replacing decks, which is only really possible in a cable-stayed bridge by reducing the structure back to the towers.

Both forms of construction are valid. The way forward for the construction of suspension bridges using galvanized high-tensile wire to form the cables has to be the installation of a dehumidification system as part of the cable installation. If this is done, and the dehumidification system is maintained and operated throughout the life of the structure, this should help to ensure that there is no deterioration throughout the life of the bridge.

Confidence in the use of the suspension bridge as a suitable form of structure, especially for long-span bridges, will only return if the service life of the main cables can be guaranteed to be the same as that of the bridge as a whole.

The retrofitting of dehumidification systems on existing main cables is strongly recommended. There can be no absolute guarantees given at present that it will prevent all future wire breaks. However, it is hoped that it will prevent further corrosion developing and future deterioration of the strength of the cables.

The assessment of the residual strength of the main cables of suspension bridges is challenging. It can only be established by carrying out an internal inspection of the main cable to determine levels of deterioration. There are no nondestructive test methods or monitoring methods currently available that can determine the strength of a main cable. It is also recommended that an acoustic monitoring system be fitted on cables where deterioration has occurred, both as an early warning system and as an aid to the selection of areas in the cable to inspect. However, further research work needs to be carried out to validate the results of acoustic monitoring systems.

There are now two differing published methods [1,2] of determining the strength of the main cables of suspension bridges. However, the methods set out in NCHRP 534 [1] Report have been used on most inspections carried out to date. It has been some time since NCHRP Report 534 [1] was written. The practical advice within the guide is invaluable. However, perhaps 10 years after publication, and with the benefit of the knowledge gained from its continual practical use during that time, a review of the guide would be beneficial. The increasing application of such a review could include the recommendations regarding the scope and frequency of inspections on bridges where dehumidification and acoustic monitoring systems have been installed.

REFERENCES

1. Mayrbaurl, R., and Camo, S. 2004. *Guidelines for Inspection and Strength Evaluation of Suspension Bridge Parallel Wire Cables*. NCHRP Report 534. Washington, DC: Transportation Research Board.

2. Mahmoud, K. 2011. *BTC Method for Evaluation of Remaining Strength and Service Life of Bridge Cables.* NYSDOT Report C-07-11. Washington, DC: New York State Department of Transport, New York State Bridge Authority, in cooperation with U.S Department of Transportation Federal Highway Administration.

3. Chavel, B.W., and Leshko, B.J. 2012. *Primer for the Inspection and Strength Evaluation of Suspension Bridge Cables.* FHWA-IF-11-045. Washington, DC: U.S Department of Transportation, Federal Highway Administration.

4. Young, J., Lynch, M., Lambert, P., and Fisher, J. 2008. *Assessment of the Suspension Cables of the Severn Bridge, UK, Creating and Renewing Urban Structures—Tall Buildings, Bridges and Infrastructure.* 17th Congress Report of IABSE. Chicago: IABSE.

5. Hossain, I. 2008. Safety Evaluation of a Suspension Bridge with Degraded Cables. Presented at the 6th International Cable Supported Bridge Operators' Conference, Takamatsu, Japan.

6. Mayrbaurl, R.M., and Camo, S. 2002. Strength and Reliability of Corroded Wire Cables. Presented at the 3rd International Cable Supported Bridge Operators' Conference, Awaji Island, Japan.

7. Hopwood, T., II. 1998. Ohio River Suspension Bridges: An Inspection Report. Presented at the Workshop on Safety Appraisal of Suspension Bridge Main Cables, Newark, NJ.

8. Service, T., McKrell, T., Mayrbaurl, R., Latanision, R., and Paskova, T. 2004. Embrittlement and Cracking of Cold Drawn High-Strength Bridge Cable Wire. Presented at the 4th International Cable Supported Bridge Operators' Conference, Copenhagen, Denmark.

9. FHWA. 2004. National Bridge Inspection Standards. 23 CFR Part 650. *Federal Register*, 69(239).

10. New York State Department of Transportation. 1997. *Bridge Inspection Manual.* Albany: New York State Department of Transportation. With updates through 2006.

11. Alampalli, S. 2014. Designing bridges for inspectability and maintainability. In *Maintenance and Safety of Aging Infrastructure*, ed. D. Frangopol and Y. Tsompanakis. Boca Raton, FL: CRC Press.

Chapter 5

Corrosion of Main Cables in Suspension Bridges

William J. Moreau

CONTENTS

5.1 INTRODUCTION

The corrosion of main cables in suspension bridges has been studied exten-sively over the past few decades. The fact that there has not been a failure or collapse of a large suspension bridge due to the effects of corrosion has made the study of this issue an uphill battle for many. The Waldo-Hancock Bridge (Figure 5.1) in Maine may have been the closest structure to a col-lapse near the turn of the twenty-first century and was strengthened imme-diately upon discovery of the extensive loss of cable strength. Figure 5.1 shows the original cables of the bridge in white, the temporary supplemen-tal cables above, and debris from broken wires in the center of the photo. The bridge was replaced with a cable-stayed bridge a short time later.

5.2 ISSUE

Wires comprising the main cables of suspension bridges are inaccessible to a general inspection, and thus many bridge owners have taken an "out of sight, out of mind" strategy in evaluating their condition. Even a carefully planned invasive inspection typically reveals only a very small percentage of the cable wires for condition evaluation. Figure 5.2 shows a typical intrusive inspec-tion of the main cables of the Bear Mountain Bridge in Fort Montgomery,

Figure 5.1 Waldo-Hancock Bridge, Maine.

Figure 5.2 Typical intrusive inspection of main cables. (From Parsons Engineering, Inspection and Condition Evaluation of the Suspension Cables for the Bear Mountain Bridge, Parsons Engineering, New York.)

New York. This figure also shows the interior broken wire pulled out for inspection and the dried condition of red lead paste. The Bear Mountain Bridge, for instance, has approximately 365 ft^2 of wire surface per lineal foot of cable. Multiplying this figure times the length of each of the two main cables produces nearly 1.9 million ft^2 of wire surface, with corrosion potential. Evaluating this potential and classifying the actual condition of these wires becomes a nearly impossible task. Keeping in mind that the status of corrosion may change over time creates a monumental effort to maintain a current reliable strength assessment for safe operation of the bridge.

Figure 5.3 Brittle wire failure due to hydrogen-assisted degradation. Note lack of corrosion. (From Modjeski & Masters, Mid-Hudson Bridge Main Cable Investigation,
Phase III, Modjeski & Masters, Harrisburg, PA, 1991.)

Water and humidity are the catalysts that drive the most significant
mechanisms for wire degradation. Basic surface corrosion or oxidation
can deplete the conventional zinc coating, which is designed to ward off
corrosion and can lead to loss of wire cross section and strength. But more
importantly, it has been determined that the atomic hydrogen, released
during the oxidation of the zinc, has a more significant chemical reaction,
causing embrittlement of the high-strength carbon steel of the wires themselves (Figure 5.3). The loss of ductility created when high-strength wires
are exposed to atomic hydrogen allows residual forces within the wires to
become stress risers with the potential to form cracks. Cracked wires induce
higher stress into the remaining section of the wire, increasing the potential for crack growth and eventual wire failure. Understanding the original
wire manufacturing techniques explains the high level of and numerous
opportunities for high residual stress to be "built in" to the high-strength
bridge wire from its conception; adding into this the cable-spinning process yields an almost infinite combination of events that can produce
uneven stress levels within the wires, which then have the potential to
form cracks during decades of service loads and through environmentally
induced degradation.

5.3 CABLE CONSTRUCTION METHODS

Cable construction methods also may have contributed to the long-term
corrosion potential of the main cables. In order to speed construction,
Washington Roebling [4] invented a new method to temporarily bind the

parallel wires into strands during the air-spinning operation (Figure 5.4). Unfortunately, he used untreated steel straps (Figure 5.5) that were to be removed during the cable compaction operation, but some straps were left in place, creating a cathodic reaction between the galvanized wires and the bare steel strap. Not only did this cathodic corrosion cell create a concentration of corrosion activity, but it also supported the release of atomic hydrogen through the corrosion process that embrittled adjacent wires.

Air spinning of the main cable wires (Figure 5.6) was the most common construction procedure for building suspension bridges in the 1920s and 1930s. It was not uncommon for the live wire to come off the spinning wheel, while it was being pulled across the span. Diaries from the Bear Mountain Bridge revealed this problem and the eventual solution implemented there. When the wire did come off the wheel and form a kink, halting the spinning progress, the wire was simply hammered out with a wooden mallet and placed back upon the wheel, and spinning resumed. Extremely high residual stresses were undoubtedly created in this operation.

The manufacture of high-strength bridge wire can also contribute to buildup of high residual stress levels within the wires. The process of cold drawing the wire, running it through an acid bath for cleaning, and then running it through the molten zinc bath required the wire to be pulled through a series of relatively small-diameter pulleys. Upon completion of

Figure 5.4 Bare steel straps intended to temporarily bind wires during original cable spinning.

Figure 5.5 Temporary steel strap from Roebling's manual of parallel wire suspension bridge construction. (From John A. Roebling's Sons, *Construction of Parallel Wire Cables for Suspension Bridges,* John A. Roebling's Sons, Bethlehem, PA, 1925.)

Figure 5.6 Original cable spinning of Bear Mountain Bridge. Note the 6 ft diameter coils of bridge wire in the background. (From John A. Roebling's Sons, *Construction of Parallel Wire Cables for Suspension Bridges*, John A. Roebling's Sons, Bethlehem, PA, 1925.)

the manufacturing, the wire was loaded onto 36 in. diameter reels for shipping. Wires removed from the Bear Mountain and Mid-Hudson Bridges for laboratory testing displayed a 54–60 in. diameter residual curvature when left unloaded (Figure 5.7). Calculating the stress required to straighten these wires results in a stress level of 50–70 ksi. Adding this residual stress to the in-service level creates a stress well above the ultimate capacity of the wire. Cracking is certain to occur at these stress levels. Stress corrosion cracking is accelerated under high-stress levels.

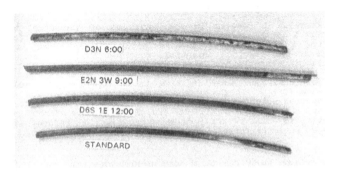

Figure 5.7 Residual curvature of wire samples.

5.4 ENVIRONMENT

Sulfur is another catalyst that can combine with other elements to create hydrosulfuric acid. Sulfur, chlorides, and nitrates have all been detected in the main cable corrosion products at the Mid-Hudson Bridge, Poughkeepsie, New York, and may have been deposited early in the life of the cable due to high concentrations of coal-burning smog, prevalent in the Hudson Valley during the original bridge construction. The Mid-Hudson Bridge cables were spun in the fall and did not receive the wrapping wire until the following summer. Snowfall over the winter may have collected and deposited pollution from the air onto and into the unprotected cables.

The preceding section of a corrosion study performed by PSG Corrosion Engineering [2] has been reprinted here as a brief discussion of the potential chemical reactions that can create a deleterious environment within main cables of suspension bridges (Figure 5.8). There is a subtle difference

PSG Corrosion Engineering   28
Mid-Hudson Bridge Phase III Study

Hydrogen Testing

HYDROGEN ANALYSIS

  Wire breaks on other bridge cables have been attributed to hydrogen embrittlement. The role of hydrogen on the cracking found in the Mid-Hudson Bridge wires had not been established previous to this study.

  It is difficult to distinguish between the two forms of environmentally induced cracking (EIC) affecting the cable wires, stress corrosion cracking and hydrogen embrittlement, because they have a similar appearance. Sometimes both can occur. References in the literature are found using the term hydrogen assisted stress corrosion cracking. There are several mechanisms by which hydrogen can damage metals (3).   The mechanism most likely affecting the bridge wire is one where atomic hydrogen enters the steel and causes a decrease in the energy necessary for a crack to propagate. The atomic hydrogen can be formed from a corrosion reaction when the solution chemistry is favorable, as follows:

    (1)   Iron -----> Iron ions + electrons (anode)

    (2)   Hydrogen ions + electrons -----> Hydrogen gas (cathode)

    (3)   Water + Oxygen + electrons -----> Hydroxide (cathode)

  Reaction 1 results in the corrosion of the steel wire. Reaction 3 is the normal opposing reaction at cathodic surfaces; however, if the pH (acidity/alkalinity) of the water is low enough at the reaction potential, reaction 2 will occur. For example, in a neutral solution with a pH of 7.0, atomic hydrogen will be formed if the corrosion reaction potential is -730 millivolt (mV) or more negative to a copper-copper sulfate reference (CSE). If the corrosion reaction potential for steel in a neutral solution is -500 mV CSE hydrogen will not be formed. However, if the pH drops to 4.0, hydrogen is formed at a potential of -553 mV (CSE) or more negative; therefore, atomic hydrogen would be formed in our example reaction.

Figure 5.8 Corrosion study results of suspension bridge main cables. (From PSG Corrosion Engineering, Corrosion Investigation of the Main Cables for the Mid-Hudson Bridge, PSG Corrosion Engineering, West Chester, PA, 1992.)

between hydrogen embrittlement and stress corrosion cracking. But to the bridge owner, the most important issue to understand is that active corrosion of galvanized high-strength bridge wires can release enough atomic hydrogen to embrittle the steel wires and reduce their ductility, which may lead to cracking and eventual wire failure.

Testing corrosion product from the Mid-Hudson Bridge did *not* confirm a theory that some form of bacteria may be feeding on the linseed oil residue within the cable and creating hydrogen as a byproduct. Researchers from the Massachusetts Institute of Technology (MIT) were retained to evaluate the "black spots" noticed during the cable inspections from the mid-1980s and determine if there were any organic organisms or evidence of bacteria that may have contributed hydrogen to the microenvironment of the bridge cable wires (Figure 5.9). No evidence was found to support the theory that the hydrogen-based wire deterioration was accelerated or caused by microorganisms or bacteria.

The corrosion evident on a small percentage of the Mid-Hudson Bridge wires consisted of black spots and corrosion pits. Many of the preexisting cracked wires located during the intrusive cable inspection of the 1990s revealed brittle wire failure beginning from either of these two flaws. A scanning electron microscope (SEM) was used in an attempt to determine the mechanics of the wire failures (Figures 5.10 through 5.14). This procedure found that many of the cracks in the wires had formed very early in the life of the bridge wire and were corroded to the point where evaluation of the intergranular structure was not possible. Sample wires were also tested to tensile failure, creating fresh fracture surfaces. Many wires were found to have lost their ductility, even though they retained their original load capacity. Failure points of the low-ductility wires, during the tensile

Figure 5.9 MIT technicians testing for hydrogen-producing bacteria at the Mid-Hudson Bridge.

MODJESKI AND MASTERS, INC.

Hudson Bridge
Cable Investigation (Phase III)   I-24

raph I-23 - Work Area E1N. View of two corrosion cavities on a wire 6
           inches below the perimeter in the 12:00 o'clock opening.
           These cavities were located on the portion of the cable
           under the removed cable band. The largest cavity was 40
           mils deep.

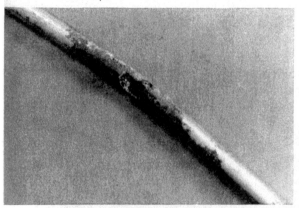

           Close-up view of 30 mil deep corrosion cavity found on a
           broken wire removed from Work Area D6S in the 12:00 o'clock
           opening 5 inches below the perimeter. Note typical
           localized depletion of zinc and steel corrosion.

Figure 5.10 Condition of the main cables at the Mid-Hudson Bridge. (From Modjeski &
           Masters, Mid-Hudson Bridge Main Cable Investigation, Phase III, Modjeski &
           Masters, Harrisburg, PA, 1991.)

tests, demonstrated stepped cracks or diagonal crack failure surfaces. Very
little necking down of the wire diameter occurred at these points of failure,
and all cracks were initiated from the inside surface of the residual wire
curvature created during the manufacturing process.

Mid-Hudson Bridge
Main Cable Investigation (Phase III)   II-15

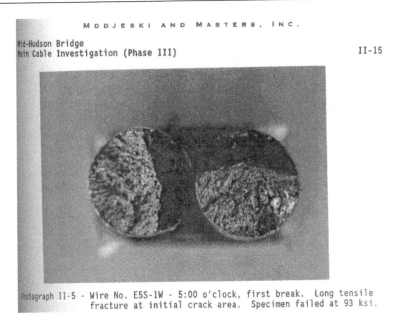

Photograph II-5 - Wire No. E5S-1W - 5:00 o'clock, first break. Long tensile
fracture at initial crack area. Specimen failed at 93 ksi.

Figure 5.11 Condition of the main cables at the Mid-Hudson Bridge showing fracture. (From Modjeski & Masters, Mid-Hudson Bridge Main Cable Investigation, Phase III, Modjeski & Masters, Harrisburg, PA, 1991.)

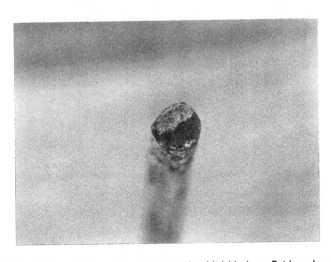

Figure 5.12 Condition of the main cables at the Mid-Hudson Bridge showing fracture surface. (From Modjeski & Masters, Mid-Hudson Bridge Main Cable Investigation, Phase III, Modjeski & Masters, Harrisburg, PA, 1991.)

PSG Corrosion Engineering
Mid-Hudson Bridge Phase III Study 23

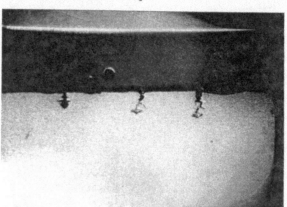

Figure 10. Sample from E3S-13E. Cross section of cracks. View is perpendicular to reel set curvature (side view).

Figure 5.13 Main cable condition at the Mid-Hudson Bridge showing crack cross section. (From Modjeski & Masters, Mid-Hudson Bridge Main Cable Investigation, Phase III, Modjeski & Masters, Harrisburg, PA, 1991.)

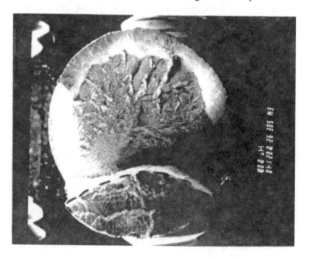

Figure 5.14 Condition of the main cables at the Mid-Hudson Bridge.

5.5 CORROSION MITIGATION

These corrosion mechanisms, described in the section above, are probable causes of the wire degradation witnessed on the cables in the Hudson River Valley, but do not represent the total list of cause-and-effect relationships driving the corrosion within suspension bridge cables. One obvious mitigation measure is to reduce the ingress of water and humidity into the main cables. Strategies applied to date include the following:

- Improved paint or coating systems that have high elongation properties, more likely to resist the formation of cracking of the outer layer of protection, for example, waterborne acrylic paint.
- Preformed membrane-wrapping systems, including neoprene, fiberglass, and polyester tape (Figures 5.15 and 5.16).
- Replacement of cable band caulking on a regular basis with durable ductile materials such as polysulfide caulk.

Figure 5.15 Cable wrap system used in Bear Mountain Bridge around 1995.

Figure 5.16 Neoprene cable wrap on I-74 bridges over the Mississippi River, Illinois to Iowa.

- Removing and replacing the seizing or wrapping wire of the main cables while using modern zinc paste or epoxy paste, instead of the conventional red lead paste (Figures 5.17 through 5.19). Modern materials have a longer service life, by remaining plastic for many years, than traditional red lead paste, which can dry out in 5 years or less. This procedure is generally accomplished during an invasive cable inspection and may not be cost-effective as a stand-alone mitigation measure.

Figure 5.17 Early use of Grignard Complex 2C zinc paste at Wheeling Suspension Bridge.

Figure 5.18 Zinc paste applied to cables of the Wheeling Bridge in West Virginia prior to wrapping.

Figure 5.19 Elettrometall as an alternate substitute for red lead paste was also evaluated at Bear Mountain Bridge.

- Larger scale mitigation measures have included unwrapping, wedging, and oiling the main cables with a non-petroleum-based oil to protect the cables from the effects of moisture and water. Original cable fabrication during the early twentieth century often used linseed oil to assist in the cable compaction effort (Figure 5.20).
- Dehumidification of main cables.

Oiling was thought to assist in the seating of the wire strands into a circular shape prior to cable compaction. Cable inspections have revealed that dried linseed oil deposits have provided an increased level of protection against surface corrosion. Early corrosion mitigation of the Williamsburg

Figure 5.20 Cable-oiling Mid-Hudson Bridge around 1994 at an application rate of 1 gal/ft of cable length.

Bridge in New York City was performed using fish oil, and more recent efforts have utilized a double-boiled linseed oil with chemical drier additives that combine with any free moisture previously deposited within the cable to produce a nonreactive compound. Some experimental work has been done using powdered reagents that also combine with any free moisture in the cables to reduce or eliminate the corrosion potential. The application of powdered reagents has not been proven to be distributed well enough to convince the author of its reliability.

The cables of the Mid-Hudson Bridge and the Bear Mountain Bridge were unwrapped, wedged, inspected, oiled, and rewrapped with conventional seizing wire, and then painted with waterborne acrylic paint. The Mid-Hudson restoration work was performed in the early 1990s, prior to the development of the zinc paste substitute, while the Bear Mountain Bridge work was performed in 1999–2000, and the Grignard Complex 2C zinc paste was used below the wrapping wire. Both bridges have received follow-up inspections where random panels have been again unwrapped, wedged, inspected, oiled, and rewrapped.

Three cycles of inspections on the Mid-Hudson Bridge, every 5 years, have resulted in no measurable increase in cable degradation. Two inspection cycles, every 5 years, have been performed at the Bear Mountain Bridge, and again, no measurable loss of strength or section was detected. While oiling a main cable is expensive and the mitigation effects do have limited performance life, more than 20 years of mitigation has been achieved using this method, and it is estimated that the useful effects of this process will endure for at least 35 years. The zinc paste has outperformed the red lead paste and is the paste of preference for all ongoing cable rewrapping work.

The most recent application of technology for cable corrosion mitigation has been the development of main cable dehumidification systems that simply blow dry air into the main cable and allow for relief ports where moist air can be exhausted. Most dry-air injections systems (Figure 5.21)

Figure 5.21 Dry-air cable injection system at Severn River Bridge in Wales. Note the injection sleeve with an inspection port.

also include a new flexible cable wrap to seal the outer surface of the cable, and care is generally taken to maintain a low pressure so that the wrap is not damaged by the internal pressure from the injection process. This technology is the subject of a separate chapter of this book, and no further discussion of the pros and cons will be made here.

Recent research (2014) performed at Columbia University has demonstrated that even parallel wire suspension bridge main cables that had previously been oiled can still be effectively dehumidified with conventional dry-air injection. Spacing between injection ports should be less than 200 m, with injection pressures around 35 Pa. The objective is to see the humidity of the exhaust air drop to 45% or below.

5.6 BEST PRACTICES

Cable bands are generally formed in two halves and bolted together at the seams located along the top and bottom of the cable. Some bridges, including the Wurtz Street Bridge in New York (Figure 5.22) and the Severn River Bridge in the UK, have their seams located along the sides of the cable. For bridges with their seams located along the top and bottom of the cable, a short length of caulk is commonly removed from the bottom groove, which may allow some free moisture to drain from the main cable. This preventative measure is recommended as a best practice for all suspension bridges with vertically mounted cable bands and without cable dehumidification.

Figure 5.22 Wurtz Street Bridge in New York State. Note the horizontal split between the cable bands, making a drain opening in the cable seam ineffective.

Cable bands must be sealed for the dry-air injection method of main cable dehumidification to be effective.

Another point of possible moisture intrusion may be found by examining the tower saddles at the tower tops. Many suspension bridges were designed with ornamental hoods over the tower saddles, including the Humber Bridge in England, the Mid-Hudson Bridge, and the Triborough Bridge in New York City. These bridges had vents built into the ornamental hoods and included access doors for inspection purposes. Water was observed in the main cable saddles of the Triborough Bridge (Figure 5.23) in the mid-1980s, and a saddle cover was fabricated to protect the main cable from the effects of the weather-driven water intrusion. The tower tops of the Humber Bridge (Figure 5.24) were dehumidified in the early 1990s, and the Mid-Hudson Bridge tower hoods (Figure 5.25) were sealed and dehumidified in 2005 to eliminate the potential of condensation creating free moisture that could enter the main cable at the unprotected saddle. Main cable inspections performed after the Mid-Hudson tower top dehumidification retrofit reported a significant improvement over previous main cable inspections with respect to the presence of small pockets of free moisture.

The Bear Mountain Bridge and the Golden Gate Bridge in San Francisco (Figure 5.26) do not have any ornamental hood or volume of air over the tower saddle and, accordingly, did not receive tower top dehumidification. Each bridge has unique details that should be inspected for any and all opportunities for water, moisture, or condensation to enter the main cable, thereby initiating wire corrosion potential.

Figure 5.23 Triborough Bridge tower top. Note the access door above the main cable and the size of the chamber above.

Figure 5.24 Humber Bridge tower top.

Figure 5.25 Mid-Hudson Bridge tower hood over cable saddle. Note the dehumidification duct.

By contrast, the Royal Gorge Bridge (Figures 5.27 and 5.28) in Colorado was constructed without the circumferential seizing wire wrapping. A personal observation was made of this suspension bridge cable by the author in 2001, and no evidence of corrosion was detected on the surface. Unfortunately, no information is available regarding the condition of the interior wires. Colorado has an extremely dry environment, and

Figure 5.26 Golden Gate tower and cable saddle. Note the lack of ornamental hood, reducing the potential for condensation to form.

Figure 5.27 Royal Gorge Bridge.

apparently, even rain does not create a moist environment long enough to allow corrosion to commence.

5.7 SUMMARY

Corrosion in main cables of suspension bridges can be driven by any number of catalysts. Many are site specific, depending on the details of construction, and some are more generic and appear common in bridges that share

Figure 5.28 Royal Gorge Bridge. Note the lack of wrapping wire or corrosion in the original construction in 1929.

similar environmental climates. The one most common corrosion driver is moisture that enters the cable and remains over time, allowing chemical degradation processes to develop and degrade the galvanized coatings and affecting the ductility of the primary bridge wire.

Accordingly, inspection methods should include observations that look for opportunities for moisture to enter the cable, as well as any signs that corrosion may be ongoing within the cable, for example, stains, bulges in the cable shape, gaps in wrapping wire, or caulking around cable bands. Also, the lack of an opening in the caulk at the bottom groove of cable bands may not allow any moisture that has entered the cable to escape. Tower hoods should be investigated for creating the potential for condensation to enter at the cable saddle, and the cable splay areas, within the anchorages, should be investigated for signs of moisture or corrosion staining that may drain down along the primary wires.

If the bridge owner chooses to engage in an intrusive cable inspection, any free moisture should be collected for chemical analysis to assist in developing a more in-depth understanding of the corrosion process that may be underway.

REFERENCES

1. Modjeski & Masters. 1991. Mid-Hudson Bridge main cable investigation, Phase III. Harrisburg, PA: Modjeski & Masters.
2. PSG Corrosion Engineering. 1992. Corrosion investigation of the main cables for the Mid-Hudson Bridge. West Chester, PA: PSG Corrosion Engineering.

3. Parsons Engineering. Inspection and condition evaluation of the suspension cables for the Bear Mountain Bridge. New York: Parsons Engineering.
4. John A. Roebling's Sons. 1925. *Construction of Parallel Wire Cables for Suspension Bridges*. Bethlehem, PA: John A. Roebling's Sons.

Chapter 6

Suspension Bridge Cable and Suspender Rope Maintenance

Daniel G. Faust, Mark Bulmer, Beverley Urbans, and David Wilkinson

CONTENTS

6.1 CORE ELEMENTS OF THE SUSPENSION SYSTEM

Suspension bridges are constructed where the longest spans are required, and it is not viable to construct intermediate piers. They maximize the use of tension; therefore, an efficient, yet robust, suspension system is a key component. The main cables are essentially catenaries, prestressed by dead weight applied via the suspenders and cable bands and tied at the ends to anchorages, which are typically either rock anchors or large gravity structures.

The main cables are comprised of either a compacted bundle of high-tensile steel parallel wires or (for shorter spans) a group of helical strands. Suspenders are wire ropes or strands connected to the main cables via cable bands, which are tightly clamped on to the cable by tensioned bolts or screwed rods. Suspenders are usually connected to the deck at an equal spacing of between 30 and 60 ft, depending on loading requirements. A smaller spacing of 30 ft can be beneficial where the bridge carries rail loading.

The span arrangement of a suspension bridge depends on the topography and geology of the site. Typically, the ratio of the suspended side to main span will be in the 0.3–0.4 range, with ratios of main span cable sag to a span of 1/8–1/11. Suspension bridges often have side towers or cable bents adjacent to anchorages where the main cable passes through a saddle that diverts the cable from the deck level down to the anchorage at the ground or water level. The splay or anchor span within the anchorage is the area where the strands diverge to their respective anchor points between the splay saddle and the anchor face.

6.2 FABRICATION AND CONSTRUCTION
  CONSIDERATIONS

Suspension bridge spans have grown to over 6500 ft with the construction of the Akashi Bridge (opened in 1998) and the planned 10,827 ft span Messina Bridge now under design. Higher-strength wire is making it possible to span further and save weight, but caution must be exercised, as higher-strength steels generally have lower ductility and defect tolerance and, consequently, decreased durability. This puts much greater emphasis on the need for effective and reliable cable preservation strategies that emphasize monitoring and corrosion.

As suspension bridge span lengths have increased over the years, the factor of safety that main cables are designed to has decreased. For example, the allowable tensile strength factor of safety for the main cables of the 1750 ft span Benjamin Franklin Bridge in Philadelphia (opened in 1926) was 4.2, while the 6532 ft span Akashi Kaikyo Bridge (opened in 1998) was designed with a factor of safety of 2.2.

The primary factors leading to decreasing safety margins are improvements in high-strength steel wire manufacturing processes, which provide greater confidence in wire quality, and an increased understanding of suspension bridge behavior. A less obvious reason is that as main spans increase, the proportion of main cable stress attributable to dead load increases, and the proportion of fluctuating live load decreases. For example, in the main cables of the Akashi Kaikyo Bridge, approximately 91% of the tensile force is attributable to dead load, whereas this figure would typically be closer to 80% for a 3000 ft span.

6.2.1 Main Cable Design Considerations

The majority of suspension bridge main cables are formed from thousands of parallel high-strength galvanized steel wires, typically around 0.2 in. in diameter. Helical strands are sometimes used for smaller-span suspension bridges. In a large-diameter suspension bridge cable, the number of parallel wires in a single cable can be tens of thousands.

The selection of wire diameter, number of wires per strand, and strand pattern is determined by the cable area required to support the bridge. Ideally, the number and size of wires will achieve a hexagonal strand cross section, with the number of strands configured to achieve a hexagonal cable. This hexagonal arrangement allows the strands to be packed neatly together in the uncompacted cable and minimizes the movement of wires when the strands are compacted into a circular cross section. This in turn minimizes the number of crossing wires in the compacted cable; the presence of crossing wires can lead to degradation of wire galvanizing and corrosion pits at crossing locations. Figure 6.1 shows the cross section of the 127-wire strands in a replica section of the Akashi Kaikyo Bridge main cable.

Galvanized high-strength steel wire with a tensile strength of 160 ksi was used as early as 1883 in the construction of the 1595 ft main span on the Brooklyn Bridge. With the construction of the 3500 ft span George Washington Bridge (opened in 1931), wire strength had increased significantly to 220 ksi. For the next 60 years, wire strength increased only slightly to 228 ksi, with the construction of the 3524 ft (1074 m) span Bosphorus Bridge, completed in 1973. The next significant increase in wire strength was motivated by a preference to construct the 6532 ft span Akashi Kaikyo Bridge using large-diameter single cables. The Akashi Kaikyo Bridge's main cable wires have a tensile strength of 260 ksi, which was achieved

Figure 6.1 Replica section of Akashi Kaikyo Bridge main cable showing 127-wire PPWSs.

by increasing silicon content (rather than carbon, which would reduce toughness).

Caution must be exercised when specifying very high-strength wire, as experience of its use is starting to reveal that in some cases, it has greater susceptibility to failure from small defects and surface imperfections. Higher-baseline stress in a stronger but less ductile material generally reduces the critical defect size, which potentially reduces durability and increases the risk of premature wire breaks.

6.2.2 Wire Fabrication

Main cable wire is manufactured from a 0.4 in. diameter high carbon (0.8%) steel rod, which is repeatedly cold-drawn through dies to achieve the required diameter, mechanical strength, and toughness properties. The wire is then annealed, pickled, hot-dip-galvanized, and coiled in a continuous process. Hot-dip galvanizing provides corrosion protection and increases ductility, but at the expense of a slight reduction in wire strength of approximately 30 ksi.

Wire is typically manufactured with a 5 ft cast diameter, which, even after many years of being pulled straight under the load in the main cable of an existing suspension bridge, will remain in the wire. This straightening of wires manufactured to a relatively tight (5 ft) cast diameter creates weak spots in the wire, as evidenced by cracks observed in main cable wire samples removed from existing cables. These cracks consistently initiate at the inside of the cast diameter, that is, at the most highly stressed part of the wire cross section when under load. The growth of these cracks is driven by the microenvironment within the cable (which contains water, zinc, iron, and contaminants), which generates hydrogen atoms; propagation of these cracks is aided by the tensile stress in the wire.

Figure 6.2 shows a cracked wire observed in a main cable and the original cast diameter of sample wire. The wire is noticeably cusping where the crack has formed, demonstrating that the crack, as always, initiated at the inside of the cast diameter.

An important development in improving durability of high-strength main cable wire is to maximize cast diameter by making it as straight as possible, thus significantly reducing stress at the inside of the cast diameter. It is believed that wire manufactured with a cast diameter of up to 65 ft is

Figure 6.2 Cracked wire in a main cable noticeably cusping where a crack has formed and original cast diameter of cable wire.

now achievable, which has a straight appearance, as opposed to the pronounced curvature of wire with the typical 5 ft cast diameter.

6.2.3 Main Cable Construction Considerations

Historically, two processes have been used to construct parallel wire main cables: aerial spinning and preformed parallel wire strand (PPWS).

Both processes involve erecting thousands of miles of wire in a controlled manner at great height, using temporary catwalks for access, in a harsh environment of extreme wind, rain, and temperature variations. The exposure to the elements and mechanical abrasion that occurs due to the erection process inevitably damages protective galvanizing, which leads to premature corrosion.

6.2.3.1 Aerial Spinning

The traditional aerial spinning process involves drawing loops of wire across the bridge using spinning wheels mounted on a tramway system that pass back and forth over the length of the suspended spans. The process was invented by French engineers Charles Bender and Louis Vicat in 1820, and was improved to achieve greater efficiency by J.A. Roebling for the construction of the Brooklyn Bridge. The process involves geometric wire-by-wire adjustment and remained largely unchanged until the early 1960s, when American engineer John Nixon had the idea of controlling wire tension rather than geometry. The controlled-tension process did not completely replace the traditional process until the late 1990s. The process is significantly less labor-intensive and causes less damage than traditional spinning due to the elimination of wire-by-wire adjustments. Figure 6.3 shows cable-spinning operations at Storebaelt East Bridge in 1996. The photo on the left shows coils of wire being reeled onto drums, which subsequently unreel the wire as it is pulled across the bridge by the spinning wheel, pictured on the right.

The wire coils used in the aerial spinning process must be spliced together using specially designed connections known as ferrules to form the continuous loops of wire that make up the main cable strands. The variation in coil size randomly distributes the splice ferrule locations along the main cables.

6.2.3.2 Preformed Parallel Wire Strand

The PPWS method involves manufacturing complete strands in a factory, which are transported to the site on large steel reels so the whole strand can be hauled across the spans on catwalk-mounted rollers. Figure 6.4 shows PPWS hauling and installation at the Akashi Kaikyo Bridge.

The strand configuration most commonly used for PPWSs is 127 wires 0.2 in. in diameter.

Figure 6.3 Cable-spinning operations at Storebaelt East Bridge in 1996.

This is a relatively small number of wires per strand in comparison to strand sizes typically used for aerial spun cables. This number of wires is a practical limit based on handling, reeling, and hauling considerations. The 127 wires also allow the strand to form a neat hexagonal cross section. The hexagonal shape is maintained by banding of the strands, usually at 5 ft intervals.

PPWSs have no splice connections within any wire over the length of the strand because each wire is cut to the same length during manufacture.

Figure 6.4 PPWS installation and hauling at the Akashi Kaikyo Bridge.

This lack of splice couplers eliminates wire fretting and permits strands to remain in a hexagonal shape, thus reducing the number of crossing wires and facilitating compaction. This increased compaction permits void ratios of 18%, as opposed to 22% for a spun cable.

PPWSs are socketed at each end, which permits attachment to anchor rods or members protruding from the face of the anchor chamber. Typically, the wires are splayed out within a conical void within the socket. The void is then filled with cold epoxy resin and a molten zinc (98%) + copper (2%) alloy to anchor the wires to the socket in a soft ductile medium, which offers cathodic corrosion protection. Epoxy alternatives have also been used for the socketing medium in place of the zinc + copper.

Clearly, both aerial spinning and PPWS erection methods can cause damage to galvanizing, resulting in corrosion "hot spots." Damage can occur during transportation and storage due to inexpert handling at the numerous points of transfer or chafing against the spinning wheels, rollers, and guides as wire is hauled across the bridge. It is incumbent upon bridge owners to understanding the extent, cause, and location of damage resulting from the erection process.

6.2.3.3 Compaction

Compaction of the erected strands into a circular cross section is carried out using purposely made compacting machines that are hauled along the cable. The compacted cable behind the machine is banded with temporary steel securing straps every 3 ft to prevent loss of compaction. If the cable bands are split vertically, the cable is compacted with a slightly smaller vertical diameter than horizontal, as this facilitates installation of the cable bands. Similarly, if the cable bands are split horizontally, the cable is compacted with the horizontal diameter smaller than the vertical.

6.2.3.4 Cable Corrosion Protection

The traditional system used to protect suspension bridge main cable wires from the environment consists of red lead or zinc paste with round wire wrapping and external paint (Figure 6.5). The wire wrapping is applied under tension at a stage when deck erection is substantially complete; this prevents a loss of tension in the wire wrapping, which would otherwise occur due to the reduced circumference of the cable under loading caused by Poisson's effect. The aforementioned wrapping machines are used to apply the wire under tension in a protective zinc or red lead paste, and the wrapped surface is then cleaned prior to painting. Wire wrapping should be applied working from the tower tops down, in order to purge water from within the cable as wrapping proceeds.

Internal cable inspections of main cables worldwide have revealed that this conventional corrosion protection system has been largely ineffective, and significant remedial actions have been necessary to deal with resultant deterioration. In response, a number of concerned bridge owners have found it prudent to flood the interior of their bridge cables with corrosion-inhibiting oil to protect the internal wires. However, there are a number of negative aspects associated with this, including the cost of removing and replacing wire wrapping, leakage of excess oil at low points, and the unknown duration of its effectiveness. Recognizing the need for a more robust corrosion protection system, most new suspension bridges are now designed with cable dehumidification systems as standard practice, to suppress the rate of corrosion. Cable dehumidification systems have also been

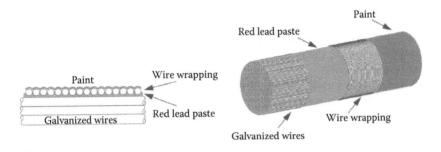

Figure 6.5 Conventional main cable corrosion protection system.

retrofitted to many existing bridges. A more detailed discussion of these and other cable preservation strategies can be found later in this chapter.

6.2.3.5 Suspenders and Cable Bands

Various suspender types have been installed over the years, each with its own production, supply, and erection processes and corrosion protection measures. The traditional configuration features four-leg wire rope suspenders, which loop over the cable bands. More recently, constructed bridges feature single or two-leg PPWS suspenders with a protective polyethylene sheath. Other configurations include open pin sockets with inclined suspenders and structural fuses for the central lock at mid–main span (Figure 6.6).

Current practice is to use epoxy filler material where the hanger enters the socket to reduce wire fretting and improve durability. Spherical bearings are also typically used on longer suspenders at quarter points where the largest transverse movements of the deck occur under wind loading. This measure also increases suspender durability.

Suspenders are typically designed to prevent progressive collapse under sudden accidental loss of two pairs of suspenders. Under normal service loading, the stress in the suspenders should typically not exceed 40% of ultimate tensile stress (UTS) on a modern suspension bridge. However, on an older suspension where the original safety factor may have been 4 or more and loss of tensile strength is extensive, resulting in a factor of safety of less than 3 (i.e., 33% of UTS), replacement of suspender ropes should be considered. Current design standards typically require that under ultimate limit state loading, which would include sudden loss of a single pair of suspenders with full traffic loading or sudden loss of two adjacent pairs of suspenders under reduced live loading, tensile stress in the suspenders should not exceed 67% of UTS, including material factors.

If cable bands are split vertically, the sole function of the horizontal bolts is to prevent slippage. The vertical bolts in horizontally split cable bands have two functions: to prevent slippage and to carry the weight of the deck. Cable bands can have a pair of grooves to carry the wire rope hangers or fin plates with pinholes to connect open sockets. Typical examples of cable bands are shown in Figure 6.7.

The length of cable bands and number of bolts in each cable band normally increase as the cable angle steepens. Cable band types are typically grouped to reduce the number of different castings required, and hence the cost of production.

Special cable bands are normally installed on the cables adjacent to the tower saddles and anchorages, which do not carry hangers. These cable bands maintain the circular shape of the main cable at the end of the transition from the hexagonal cross section in the saddles.

Cable band bolts often have a wasted shank to prevent breakage at the thread. Equal upper and lower gaps should be maintained between the

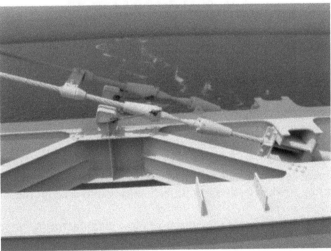

Figure 6.6 Inclined suspenders with structural fuses for the central lock at mid–main span.

two halves of the cable bands when the bolts are tightened. These gaps should be caulked with highly durable air- and watertight sealing materials to keep water out. Cable band bolts are tightened and checked to a defined sequence (and will be rechecked after deck erection).

Figure 6.7 Cable bands with PPWSs and wire rope hangers.

6.3 ROUTINE INSPECTIONS

As main cables and suspenders are the primary load-carrying elements of
the structure, it is important that their condition is known and can be
closely monitored. Bridge owners must consider which elements of the suspension system should be inspected visually on a routine basis, and what
purpose these inspections serve.

In order to assess the frequency and level of detail of routine field inspections of main cables, suspenders, and anchorages, the bridge owner should
start by using a risk assessment approach, taking into account the following:

- The age of the bridge (material grades, manufacturing standards at
  the time of construction, etc.)
- Historic data and information, such as evidence of water infiltration,
  rust staining, failure, wire breaks, paint cracking, other bridge system
  data, and so forth
- Original design information; for example, which elements are working hardest?
- Bridge loading and any changes since design
- Which elements have difficult access and therefore have not been regularly inspected and could contain hidden defects

- Yearly weather patterns (cable walking may only be possible at certain times of the year)
- Current and historic security arrangements

Routine visual inspections are an essential aspect of bridge maintenance and generally form the basis for scheduling structural maintenance activities and more detailed future inspections. Visual inspection requires careful record keeping for future comparison of inspection findings. With regard to the key elements of the suspension system, the items listed below provide an indication of what to look for when carrying out visual inspections.

6.3.1 Cable Wrapping

The condition of the main cable wrapping and paint system should be inspected for paint cracks, corrosion staining where water has seeped out, external corrosion, loose or damaged wrapping wire, and if an elastomeric wrapping system has been installed, unbonded laps or lesions. Low points of the main cables tend to be areas where water accumulates internally and are also closer to the traffic "splash zone." For these reasons, they are often found to be among the most deteriorated main cable locations. Routine inspections should also consider the potential long-term creep of the main cables and suspenders, where implications could include cable band bolt relaxation and potential slippage of cable bands (due to the Poisson effect), increase in main cable sag, cracks opening up in paint as adjacent turns of cable wire wrap move apart, and resultant water ingress caused by cracking and dislocation of wire wrapping. Figure 6.8 shows cracks in paint, and water and corrosion products seeping out of the main cable.

6.3.2 Cable Bands, Hand Strands, and Posts

It is highly unlikely that issues with the structural integrity of cable band castings will be observed, but even so, this must be checked. External corrosion due to breakdown of the paint system or deterioration in the caulking materials is a common issue that requires repair. Cable band bolts must be checked for evidence of corrosion and cracking in nuts or bolts. Bolt tensions are not normally checked as part of visual inspections. Hand strand tension must be monitored, and any change in tension along with corrosion, integrity of connections to posts, and posts to cable bands should be inspected.

6.3.3 Suspenders

Wind- and rain-induced vibration of suspenders can produce large stresses and premature failure of suspenders; therefore, early detection and installation of countermeasures, such as Stockbridge dampers or spacers for multileg

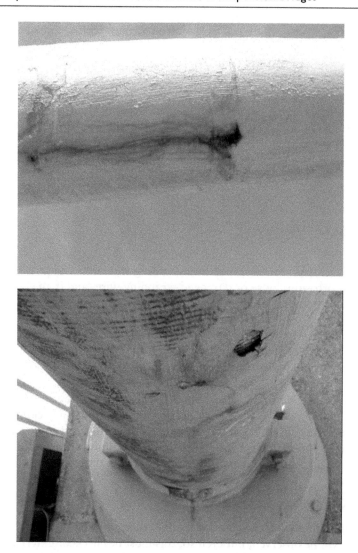

Figure 6.8 Cracks in paint, and water and corrosion products seeping out of the main cable.

suspenders, is imperative. Paint cracks, corrosion staining, broken wires, pull-out at sockets, and integrity of pinned or other socket connections are all factors that will help to establish an appropriate timescale for suspender replacement; particular attention should be paid to the bottom of the suspender ropes adjacent to sockets, as this area is susceptible to broken wires, as a result of water penetrating the protective outer layers and draining down to the socket interface (see Figure 6.9 for suspender rope socket interface).

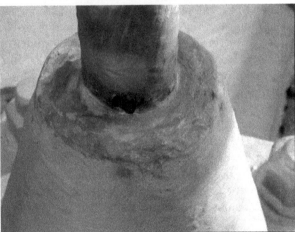

Figure 6.9 Suspender rope at the interface with a lower socket prone to high bending stresses, wire cracks, and water ingress.

6.3.4 Anchorages

The condition of splayed strands, anchor points (such as strand shoes or eyebars), and splay saddles should be accurately recorded. Particular attention should be paid to eyebars at the anchorage block interface, as moisture will often form at this point, causing "necking down" of the eyebar thickness. Any evidence of pull-out of PPWS sockets should be noted and recorded, and immediate countermeasures should be implemented. Environmental conditions, including humidity and standing water in the splay chambers, should be recorded so any necessary measures can be implemented as quickly as possible.

6.3.5 Environmental Considerations

There are numerous environmental considerations to take into account when deciding on a maintenance and inspection regime for the main cables and suspenders. It is extremely difficult to ensure that the main cable and suspenders are completely sealed. One of the main factors affecting moisture ingress into the main cables is believed to be through the effect of cyclic changes in temperature of the cable, causing expansion and contraction on a daily cycle, which leads to atmospheric moisture being sucked into the cable through small gaps in the wrapping or through caulking at cable bands. The moist air then condenses within the cable, remaining inside the wrapping, making an aggressive environment for the wires. The stretch of the main cables and suspenders causes traditional circumferential wire wrapping to separate, leading to cracks in the paint in these locations. Using an elastomeric wrap/paint and providing a positive pressure inside the main cable by installing a dehumidification system helps to alleviate this problem.

The proximity of the bridge to factories producing polluting chemicals is also an issue. Testing of water samples within the cables and suspenders will reveal if any contaminants are present. Similarly, sections of the main cables and hangers close to traffic splash zones should be prioritized for inspection due to the increased potential for chemical pollution, corrosion risk (salts), and possibly mechanical damage.

The direction of the prevailing winds (and weather) can have an impact on the condition of various elements of the bridge. Some of the suspenders may be prone to oscillations at certain wind speeds and directions, which could reduce the fatigue life of the suspenders. Several bridge owners have installed dampers on some or all of their hangers to reduce these oscillations.

Some bridges are susceptible to snow and ice accumulation on the main cables, leading to hazardous conditions for bridge users and to temporary closure in some instances. On some bridges where this is a frequent issue, snow and ice cleaners have been installed to remove the buildup and reduce the risk of having to close the bridge (e.g., Port Mann Bridge, Canada).

6.3.6 Routine Inspection Access and Other Considerations

Methods of routine visual field inspection of the main cables and suspenders are typically as follows:

- Inspection by walking at deck level to inspect the lower sections of suspenders. The underside of the cable and upper suspenders can be inspected from a distance using binoculars from the deck level.
- Inspection by walking on the main cables. This method allows close inspection of the top surfaces of the main cables. Use of mirrors on extension poles tethered to the inspector (or similar) allows inspection on the underside of the cable, cable bands, cable band bolts, and suspender sockets.
- Inspection using rope access operatives to closely inspect (within touching distance) the underside of the cable, cable bands, upper suspenders, and suspender sockets. Usually, rope access is used if there is an area flagged during the cable walk that requires closer inspection.
- Cable wires or strands in the anchorages will be inspected using permanent or temporary access provided within the anchorage chambers. If permanent access is limited, the bridge owner should consider (as a minimum) installing temporary scaffolds to inspect the usually inaccessible wires/strands to establish a benchmark condition. The level and frequency of subsequent inspections will depend on the benchmark inspection condition.
- Regular level surveys of the bridge profile at steady temperature and minimal traffic can also help to provide information for the bridge owner on sag of the spans, which could be an indication of long-term creep of the main cables and suspenders. Creep of main cables and suspenders is evident on several bridges, especially those using spiral strands or ropes where the lay of the wires can "untwist" slightly over time. This rate of creep should be monitored to check that it is gradually reducing. These surveys can be either carried out in-house or sublet to surveying specialists. If the bridge has a structural monitoring system, a hard level survey is a useful check on the system. For the level surveys to be used for comparison from year to year, it is critical that the temperature of the bridge during the survey is known.
- It is important that visual inspections record the location and type of defect, weather conditions (e.g., has there been a recent episode of heavy rain?), date of inspection, identification of the inspector(s), and good photographic records with consistent location references.

If the routine field inspections indicate that corrosion is taking place within the main cable or suspenders, more detailed investigations should be undertaken.

6.4 DETAILED INVESTIGATIONS

As previously indicated, the results of routine inspections should be used to identify the extent and frequency of more invasive and detailed internal investigations of the suspension system. These more detailed investigations are essential to establishing accurate baseline conditions, which will guide the development of a targeted maintenance strategy. National Cooperative Highway Research Program (NCHRP) Publication 534 (*Guidelines for Inspection and Strength Evaluation of Suspension Bridge Parallel Wire Cables*) provides general guidance for determining the frequency of internal cable investigations based on the age of the bridge and previously recorded wire conditions. An initial main cable investigation is highly recommended for any bridge with over 25 years of service. The most critical factor in determining the extent and frequency of internal investigations is the detection of broken wires. Understanding the cause and extent of wire breakage is critical to ensuring the long-term safety and viability of the main cable system.

6.4.1 Main Cable Investigations

Investigating the condition of main cables is challenging due to their location, size, and method of construction. As previously discussed, cables have historically been coated in paste, wire wrapping, and paint. A visual inspection of the outer surface gives very little information on the overall condition of the cable and can be misleading. It is therefore desirable to inspect the internal condition of the cable, and over the last 20 years, a technique has been developed based on driving wedges into the cable, allowing inspection of a small percentage of internal wires. This technique has been codified in the aforementioned NCHRP Publication 534, which provides guidance for establishing the baseline condition of main cables by performing a scoping study, fieldwork, laboratory testing, and strength assessment.

6.4.1.1 Scoping Study

The aim of the scoping study is to identify the panels that are to be inspected. This stage involves a review of any previous inspections, consideration of any known problem areas, and a visual inspection to identify any obvious issues, such as water blisters on the surface of the cables. Panels are typically selected to give a distribution across both cables, high and low, and in the various spans, but areas of concern may bias the inspection toward specific areas.

The NCHRP guidelines recommend that for a second inspection, when stage 4 wires have been found to a depth of more than three wires, at least 16% and preferably 20% of the panels in each cable should be inspected. If broken wires have been previously found, the recommended interval between inspections is 5 years. If a large number of broken wires have been found in the first inspection and the main cable has been found to have a significant loss in strength, a shorter interval is considered appropriate.

The suggestion that between 16% and 20% of the cable panels should be opened assumes that there has been no intervention. If a full-length acoustic monitoring system has been installed and a cable dehumidification system fully commissioned, this recommendation can be reviewed and adjusted, for example, reduced to between 4% and 5%, depending on the level of intervention since the first internal inspection.

Part of the logic within the NCHRP of inspecting so many panels is to improve the accuracy of the strength calculation, and part of this is obtained through many more sample wires being tested. Inspecting fewer panels will leave a greater uncertainty, but this will be less than for the first inspection.

The second inspection presents an opportunity to reinspect panels that were inspected during the first inspection. This is considered to be an important component of the second inspection in providing information on the deterioration rate of the cable. For example, during the Forth Road Bridge second cable inspection, two of the three panels inspected had been previously inspected. The interval between inspections was about 4 years. Of the two reinspected panels, one was found to have changed by a marginal amount, and the other had a more noticeable deterioration.

However, it should be remembered that visual inspection is a statistical sampling exercise, and a second inspection will not exactly replicate the first, as wedge lines will expose different wires/faces of wires. Therefore, it is possible that a panel could appear to be in a better condition than previously thought. There needs to be a balance between reinspecting panels previously viewed and opening new panels to gain a wider view of the overall condition of the cables.

From experience with UK bridges, the first inspection has generally produced the greatest loss in strength in the panels at mid–main span (i.e., low points). However, panels higher up the cable with a steeper inclination will have higher tensions; therefore, deterioration of these panels could be more critical due to the higher stresses. Some higher-level positions should be examined where there are other indications of deterioration (e.g., acoustic monitoring), and perhaps to improve the geographic spread of inspected panels. Other bridges have shown loss of cable strength close to the top of the cables adjacent to the towers, and others at quarter points, so unless the whole panel is unwrapped, the worst panel will not necessarily be found. When available, acoustic monitoring data, dehumidification data (e.g., are some sections of cable taking longer to dry out than others?), and other observations should be used when selecting panels. As a general guide, the following factors should be taken into account when selecting the number and location of panels to be inspected:

- The most deteriorated previous panels should be reinspected.
- Inspection of an equal number of panels on the two cables is beneficial.
- Panels with large numbers of wire breaks recorded by an acoustic monitoring system should be inspected, with the added benefit of repairs that will be carried out to broken wires found during inspection.
- Length of contiguous panels on each cable will help strength assessment calculation, as actual adjacent panel conditions can be utilized.

6.4.1.2 Fieldwork

Once the contractor has erected the cable access (scaffold, catwalks, or cable access gantry), inspection fieldwork commences at the first panel to be inspected. Depending on the schedule, the contractor may have more than one work crew in operation at the same time. Wrapping wire and protective paste are removed, and it is important that the paste is removed in its entirety to allow inspection of the outer wires.

Depending on the age of the structure and condition, broken wires are sometimes discovered upon removal of the cable-wrapping wire. Any broken wires found as the wrapping wire is removed are repaired. The repair method consists of a length of new cable wire spliced in between the two loose ends of the broken wire. The new wire is secured with a wire ferrule and turnbuckle to allow the wire to be tensioned (Figure 6.10).

Wedging of the cable is then carried out, with the wedges typically driven in at eight wedge lines around the circumference of the cable, at 12:00 on the clock face, 1:30, 3:00, and so forth. The wedge lines are opened up either individually or as sets of two or more wedge lines. It is important at this stage that the wedges are driven to the center of the cable in order to expose the full segment of the cable. Experience has demonstrated that this is more easily achieved when only one wedge line is opened up at any one time (Figures 6.11 and 6.12).

Inspection of the wires then proceeds. Each cable panel (i.e., the length between two cable bands) is typically divided up into sections to allow for ease of inspection. In accordance with NCHRP Publication 534 guidelines,

Figure 6.10 Repair of broken wires.

Figure 6.11 Main cable wedging.

all wires within the exposed wedge are graded 1–4, in accordance with the following:

- Stage 1 wire: Spots of zinc oxidation
- Stage 2 wire: Zinc oxidation of the entire surface
- Stage 3 wire: Spots of brown rust covering up to 30% of a 3–6 in. length of wire
- Stage 4 wire: Brown rust covering more than 30% of a 3–6 in. length of wire

Figure 6.12 Internal cable wire conditions.

As each wedge is opened, any broken wires that are found are repaired as per the procedure described above. Samples of wire are also taken for testing in order to provide material data for use in the later cable strength assessment. Sample wires for testing are selected by the engineer as the inspection progresses. These are chosen to give representative material strength data for the various stages of corroded wire. The wedging, inspection, repair, and sampling cycle continues until all wedges on that panel are complete. The cable is then recompacted and the protective system reinstated.

Figure 6.11 Main cable wedging.

all wires within the exposed wedge are graded 1–4, in accordance with the following:

- Stage 1 wire: Spots of zinc oxidation
- Stage 2 wire: Zinc oxidation of the entire surface
- Stage 3 wire: Spots of brown rust covering up to 30% of a 3–6 in. length of wire
- Stage 4 wire: Brown rust covering more than 30% of a 3–6 in. length of wire

Figure 6.12 Internal cable wire conditions.

As each wedge is opened, any broken wires that are found are repaired as per the procedure described above. Samples of wire are also taken for testing in order to provide material data for use in the later cable strength assessment. Sample wires for testing are selected by the engineer as the inspection progresses. These are chosen to give representative material strength data for the various stages of corroded wire. The wedging, inspection, repair, and sampling cycle continues until all wedges on that panel are complete. The cable is then recompacted and the protective system reinstated.

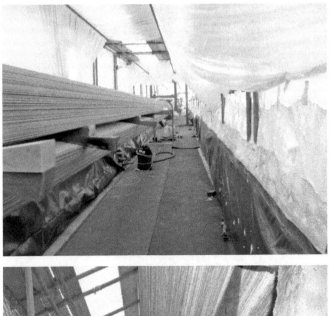

Figure 6.11 Main cable wedging.

all wires within the exposed wedge are graded 1–4, in accordance with the following:

- Stage 1 wire: Spots of zinc oxidation
- Stage 2 wire: Zinc oxidation of the entire surface
- Stage 3 wire: Spots of brown rust covering up to 30% of a 3–6 in. length of wire
- Stage 4 wire: Brown rust covering more than 30% of a 3–6 in. length of wire

Figure 6.12 Internal cable wire conditions.

As each wedge is opened, any broken wires that are found are repaired as per the procedure described above. Samples of wire are also taken for testing in order to provide material data for use in the later cable strength assessment. Sample wires for testing are selected by the engineer as the inspection progresses. These are chosen to give representative material strength data for the various stages of corroded wire. The wedging, inspection, repair, and sampling cycle continues until all wedges on that panel are complete. The cable is then recompacted and the protective system reinstated.

If a cable dehumidification system is in place during an internal cable inspection, the integrity of the dehumidification system will be compromised when the wrapping is removed for inspection. The dehumidification system is normally switched off at the beginning of a work shift, allowing cable inspection work, and then switched back on again at the end of the shift after the panels have been temporarily wrapped in plastic shrink-wrap material (or similar) and strapping to try to retain air pressure of the dehumidification system.

Given the importance of gathering cable information whenever practical, provision for further cable investigation should be made in follow-up contracts whenever possible. Further investigation would become more pressing should poor cable conditions be discovered at any stage.

6.4.1.3 Laboratory Testing

Wires removed from the cable are cut into samples, and the following data is then collected during laboratory testing:

- Tensile strength for use in cable strength assessment
- Cracked wire data across the stages of corrosion for use in cable strength assessment
- Carbon content to demonstrate correlation with tensile strength
- Zinc coating weight to demonstrate compliance or otherwise with the original bridge wire specification

6.4.1.4 Cable Strength Assessment

Once the inspection and testing data has been collated, then the process of cable strength assessment can begin using statistical modeling. NCHRP Publication 534 identifies three types of model:

1. Simplified strength model: Recommended only for use in cables with zero cracked wires or, at the discretion of the owner, cables with less than 10% cracked wires of the total number. When used on cables with cracked wires, the calculated strength of the cable may be up to 10% less than that calculated using the brittle wire model.
2. Brittle wire model: Recommended by NCHRP for use in the majority of situations.
3. Limited ductility model: The most complex of the three models and recommended for use where wire tests indicate an unusual distribution of strengths. This model requires the ultimate strain of each test specimen and a full stress–strain diagram.

The appropriate model should be selected based on the condition of the cable. The presence of cracked wires has a significant impact on the calculated strength, as does the distribution of tensile strengths, and it is therefore vital that this information is collected in a carefully controlled manner.

While the available budget of the bridge owner may affect the choice of model, it should be recognized that oversimplifying the modeling may underestimate the strength of the cable. Given the significance of the assessment and the potential decisions made based on the results, it is important that the most appropriate model is selected.

The strength assessment is concluded with an estimated factor of safety for the cable. This vital parameter will inform the bridge owner of how much of the original capacity has been lost and will feed into the overall cable management strategy. This will include planning for the next inspection, the interval of which will depend on the age of the bridge, the level of corrosion, and the number of cracked wires found.

6.4.2 Cable Band Bolts

Cable band bolts may require replacement during the life of a suspension bridge, and tensile testing of entire bolt assemblies can be used to verify the strength of the assembly. This would normally be in addition to standard testing of small sample pieces taken from the material cast. The testing rig should be capable of recording strain as well as load, in order that the stress–strain curve can be generated and assessed for material behavior. Where load–extension curves are required for determining the extension to be applied in the field, consideration should be given to taking manual measurements with a frame-based micrometer. This removes the effect of any bending of the test rig or bedding in of the bolt assembly. The mode of fracture should also be examined to verify that the material fails in the expected manner. Figure 6.13 shows the expected necking of a bolt immediately prior to failure.

During their service life, cable band bolts will relax and lose tension over time. The loads in the bolts should be checked at periodic intervals, either by physical measurement of the bolt lengths or by using ultrasonic equipment. Bolts should be retightened as required, and this should be anticipated approximately every 10 years.

6.4.3 Suspenders

Nondestructive techniques can be used for investigating the condition of suspenders. The techniques include ultrasonic testing, eddy current testing, magnetic flux leakage testing, and x-ray analysis. These forms of testing can be used to identify defects in the suspenders, for example, cracks in wires, and can be used to help assess the condition and residual life of these critical components. Much of this type of equipment has traditionally been quite bulky; however, lighter equipment is becoming available,

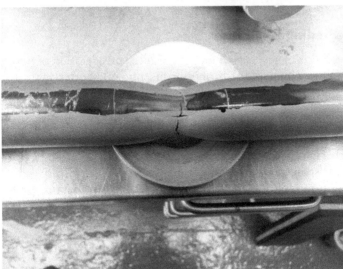

Figure 6.13 Cable band bolt failure testing.

making its use in the field easier. It should be noted that these methods do not currently permit inspection of the critical area adjacent to the socket, and effectiveness of the technique is therefore limited.

The loads in suspenders can be estimated by measuring the frequency response to excitation forces on the suspenders. Accelerometers attached to the suspenders produce a voltage output signal that is amplified, filtered, and sampled to determine the natural frequency from which the load in each suspender can be calculated. While seemingly a useful tool in estimating suspender loads and comparing their distribution around the bridge, it should be noted that the technique has been shown to work more effectively on longer hangers.

Depending upon the results of visual inspections and nondestructive testing, owners should consider removing and testing a limited number of suspender ropes. Removal of ropes will permit detailed investigation of internal suspender wire conditions and material testing. Because the process is expensive and requires replacement of removed suspenders with newly fabricated units, the number of suspender ropes selected for removal should be limited (typically three to four locations, depending upon the size of the bridge and visual conditions).

Sections of the removed suspenders should be load tested; ideally, testing will occur using the existing socket connection (to replicate actual conditions) and a fabricated socket (to replicate idealized conditions). Testing should be carried out to determine ultimate load and modulus of elasticity. Visual examination of wire should occur before and after testing. Visual examination of sockets should also occur by splitting the socket cones and examining interior wire conditions.

6.4.4 Loading Regimes

The long spans of suspension bridges require consideration of nonstandard loading when compared with more conventional shorter-span structures. Application of standard bridge loading would be inappropriate due to the intensity and total load that would be applied, and would not be representative of actual loading conditions along the loaded lengths. Bridge-specific loading is therefore more appropriate, and the challenge to the bridge owner is how this loading can be developed to provide an accurate picture of how heavily the structure is currently being loaded.

Weigh-in-motion (WIM) systems are designed to capture vehicle data such as axle weight and vehicle weight. This data can then be used to develop a bridge-specific loading, which can be used to more accurately predict stress levels within the suspension system. These systems do not require the vehicle to be stationary, but as the name suggests, the systems capture the data as the vehicles pass over them. Different types of WIM systems are available, based around the following technologies:

- Piezoelectric sensors: Load is calculated from a change in voltage as the vehicle travels over the sensor.
- Bending plate: Load is calculated from strain measurements taken from gauges bonded to plates.

- Load cell: This typically contains two scales to measure the left and right weights independently.

NCHRP Publication 683, *Protocols for Collecting and Using Traffic Data in Bridge Design*, provides a more detailed description of the pros and cons of each of these technologies, and is an excellent guide for owners in the selection of the appropriate type of system.

Once captured, the WIM data can then be used to develop a load model for the structure, and subsequently used in the detailed assessment of that structure.

While the systems have both an installation and ongoing cost of maintenance, the benefit to the owner comes from having detailed information on the traffic loads on the bridge. This provides an important tool in the long-term management of the structure, allowing the owner to monitor traffic levels and makeup over time, which can then be fed into structural assessments. The data may also be used during maintenance projects to assess the impacts of any changes to the structure. The benefit to the owner is therefore likely to compensate for the cost over the whole life of the system.

6.5 DEVELOPMENT OF A SUSPENSION SYSTEM PRESERVATION STRATEGY

The results of routine and detailed condition assessments provide a snapshot of a bridge suspension system at a particular moment in time, and will form the basis for the development of measures to enhance and preserve the system. The ultimate selection of which measures to implement should be based on a long-range strategy for extending the life of vital suspension system elements, in particular the main cable.

Proactively monitoring the condition of the suspension system is essential to developing, implementing, and modifying an effective preservation plan. Ongoing monitoring over a period of time will allow owners to identify deterioration trends and permit adjustments to preservation strategies in a targeted fashion. The most important and effective monitoring tool is the performance of detailed internal and nondestructive investigations of the main cable and suspenders on a routine basis. Ongoing inspection and strength assessment of cables will inform the bridge owner of the amount of deterioration that the cables have suffered since construction, allowing a revised factor of safety to be calculated. As each inspection is carried out, this will provide another point on the strength–time graph, and projections can be utilized to predict likely future strength loss. The rate of deterioration demonstrates how quickly the loss of strength is occurring, and how much more strength loss can be expected in future years if the deterioration is not arrested. Optimistic and pessimistic scenarios of strength loss may both be considered in the assessment process, which should also take into account all previous findings.

A diminishing factor of safety is the primary concern for bridge owners and will drive the implementation of preservation strategies. Analysis of data from around the world suggests that as the factor approaches 2 for main cables, particularly for older bridges, some form of remedial action is necessary. Ideally, such remedial measures will be implemented at the onset of deterioration in order to achieve the greatest value and extend the life of the cables for as long as possible.

Since an accurate estimation of wire breaks is the most important factor in determining cable strength, bridge owners should consider augmenting routine and detailed investigations with systems that can provide continuous health monitoring of cables and other elements. Acoustic monitoring systems have been widely used by bridge owners to detect wire breakage within main cables; the analysis of acoustic monitoring data over a length of time can be used to demonstrate whether the deterioration is continuing or has been slowed by the implementation of preservation measures such as dehumidification or oiling.

6.5.1 Planning of Preservation Measures

As indicated above, rather than considering preservation measures as discrete, short-term improvements, they should be part of a larger strategy that recognizes that wire conditions may change or new conditions may be discovered during future internal investigations or through a continuous monitoring program.

Development of capital improvements to preserve suspension system elements should be viewed in the broader context of other bridge needs. For example, the need to replace an aging bridge deck presents an opportunity to reduce dead load, thereby increasing the main cable safety factor. Similarly, replacement or enhancement of bridge systems (electrical, communication, etc.) should consider the demands of future preservations or monitoring systems.

Access and staging issues are important considerations for planning and programming of major capital improvements to the suspension system. For example, local trade agreements or statutes that promote separation of work functions will often result in access and work platforms that extend the length of the main cable in order to accommodate multiple work crews. In such cases, owners would be wise to consider maximizing the value of such access by using a system approach that incorporates improvements to all elements. In all cases, given the vital importance of the suspension system, bridge owners would be wise to institute formal processes that will evaluate the impacts of all capital improvement elements on the suspension system. Such a process will ultimately permit the development of a holistic and long-term capital improvement strategy that makes efficient use of capital funds.

6.5.2 Main Cable Preservation

Unlike most other bridge elements that have a finite and often limited expected life, main cable improvements should be guided by the principle that cable replacement is the owner's last resort. As a result, the evaluation and comparison of preservation options should consider life cycle costs associated with the option itself, not the core main cable element.

Capital improvements that promote cable preservation will generally fall into two broad categories: external and internal. Below is a brief overview of the types of improvements.

6.5.2.1 Main Cable Preservation: External Measures

External measures are intended to create a watertight seal or encapsulation of the main cable, thereby preventing water infiltration. Recognizing that water will work its way into the cable over time, such systems must include provisions for preventing water that has entered the cable from becoming trapped.

As previously discussed, the traditional "tried and true" method of protection, dating back to the Brooklyn Bridge, is a layered system consisting of an external coating applied to a circumferential wire wrapping, which in turn is placed on top of a layer of red lead paste applied directly to the main cable wires. This system has the benefit of wide familiarity and commonly used materials that can be easily maintained. However, such a system requires frequent maintenance and is ultimately ineffective in keeping moisture from penetrating the cable interior.

Due to environmental concerns with the handling and disposal of lead-based materials, oil-based zinc paste has been specified for the protection of main cables in the last few decades. Today, there are two formulations of waterproofing paste that are normally specified for U.S. bridges: Elettrometall and Grikote-Z. Elettrometall cures to a flexible rubbery state, while Grikote-Z remains pasty.

Suspension bridge cables were originally protected with the same paint system used for the steel structure. More recently, the industry has recognized that such coatings are ineffective as a waterproofing layer given the inherent flexibility of bridge cables, which promotes separation and movement of the underlying wire wrapping. As a result, water-based acrylic coatings that contain highly elastic polymers and cure to a rubbery coating are widely used today for painting suspension cables. Because of their ability to sustain up to 200% elongation without cracking or peeling, they have been successfully used for the maintenance painting of wire-wrapped cables on many existing suspension bridges and new bridges. In addition, these coatings have proven to have a long life in other applications, especially in environments where saltwater and chemical resistance are required. One such material is the proprietary coating Noxyde, manufactured originally in Belgium and now licensed for manufacture in other countries. Flexible coatings have the benefit of ease of application and maintenance; however, as with all external coating systems, failure over time will invariably occur, resulting in the need for ongoing and relatively frequent applications.

More recently, elastomeric membranes have been used as an alternative to external coating systems. The most prevalent and widely used membrane material is a product manufactured by DS Brown. The material is generally considered to be more durable than paint systems, thereby requiring less life cycle maintenance. The system includes provisions and details for creating sealed conditions at cable bands and saddle locations, and its application is less dependent upon environmental factors, which is a critical requirement in exposed conditions. Installation of Cableguard™ does require the use of trained crews and the fabrication of a specialized wrapping machine and heating blankets (Figure 6.14).

S-shaped wire wrapping is an alternative to a membrane system, and one where a wrapping wire having an S-shaped section is wound around the circumference of the cable in a spiral configuration. The hook portions of the S-shape engage with each other, creating an excellent seal that is effectively watertight. In addition, the flat surface created by the S-shape provides for better adherence of coatings than normal round wire. S-shape wire is generally considered to be a high initial cost alternative, although the greater longevity of the product may lead to lower life cycle costs when compared to other external systems. Installation requires the use of trained crews and a specialized wrapping machine.

Figure 6.14 Installation of membrane wrapping.

For a new suspension bridge, installation of S-shaped wrapping wire has become increasingly cost-effective and can be made air- and watertight with the application of a suitable external paint system. However, as a retrofit application, to replace the existing round wire wrapping with S-wire and a flexible paint system could substantially increase the cost, compared to an overwrapped elastomeric system. In addition, it is generally beneficial to retain the round wire wrapping where possible, as it provides an essential constraint to a deteriorated cable by helping to redevelop tension in any broken wires.

6.5.2.2 Main Cable Preservation: Internal Measures

Internal measures are designed to impregnate the interior of the main cable, providing protection of cable wires from within. Internal measures can and should be implemented in conjunction with external measures, although consideration to proper venting and removal of excess material must be provided.

A common internal measure is to provide a layer of waterproofing within the cable through the injection of linseed oil. This method involves unwrapping and injecting the oil at each panel point along the length of the cable, and then rewrapping and sealing the cable. The procedure uses the interstices between wires to distribute the oil. It is important to provide for drainage ports at the low point in the cable to permit excess oil to escape over time. This method has the benefit of familiarity and use of relatively common materials; however, it is also characterized by a high initial and life cycle cost, given the requirement to open and inject oil at every panel, and the need for repeated and routine applications. This method also requires that a system for collecting and disposing of excess oil be in place for months, and sometimes years, following initial application. Finally, oil that has bled through to the cable surface over time will require additional cleaning prior to the application of external coating systems (Figure 6.15).

The newest and most promising internal preservation measure is the injection of dry air into the cable in order to reduce relative humidity and effectively halt the corrosion process. Protection of steel through the control of humidity is a long proven technique dating back to the first half of the twentieth century. In simple terms, if the relative humidity can be maintained below 40%, then no corrosion will take place, as demonstrated by the graph in Figure 6.16.

The application of dehumidification for suspension bridge cables was conceived and developed in Japan for the Akashi Kaikyo Bridge, and has since been adapted as a retrofit measure on a number of suspension bridges. In particular, three bridges in the United Kingdom (Forth Road, Severn, and Humber Bridges) have been retrofit with dehumidification systems that follow the original Japanese concept, but with further development to enable an initial rapid drying, followed by reduced operation for more economical operation.

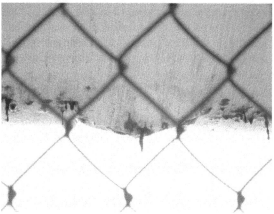

Figure 6.15 Latent effects of cable oiling.

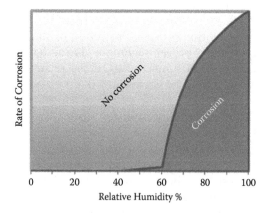

Figure 6.16 Correlation between humidity and rate of corrosion.

Dried air is produced in sealed plant rooms and blown into sealed injection points on the cables under pressure. The dried air then passes inside the cables through the small gaps between the wires, picking up moisture before being exhausted into the atmosphere, or alternatively recycled (Figure 6.17). Control and monitoring of the system is accomplished through a network of sensors and computer equipment, usually via a dedicated website.

If cable dehumidification is to be successfully achieved, the external corrosion protection system has to be as air- and watertight as possible, which

Figure 6.17 Dehumidification injection and exhaust points.

is particularly difficult to achieve at cable band locations. Various wrapping and paint systems have been used to successfully accomplish this, including the previously mentioned wrapping systems.

The location of the plant room and availability of on-site power are important considerations in designing and implementing a cost-effective cable dehumidification system. Location is dependent upon a number of factors, most notably, availability of space, adequacy and proximity of adjacent power, and ease of access. The power supply is an important consideration, as the electrical demands of the new equipment can often exceed available on-site supply. See Figure 6.18 for a typical plant room interior.

Cable dehumidification systems have an initial construction cost that is on par with cable oiling, with much of the cost attributable to accessing and sealing the cable. However, the life cycle costs of dehumidification are quite favorable compared to other methods, such as oiling, which require repeated applications. Because cable dehumidification is the only preservation strategy that actively inhibits the corrosion process by changing the internal environment within the cable, it is garnering increased interest from suspension bridge owners around the world. Ongoing maintenance of the systems is straightforward, provided that standard wrapping and sealing materials, plant, and controls are used, and the plant is well designed and easily accessible.

Figure 6.18 Dehumidification plant room interior.

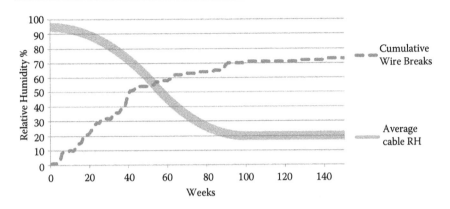

Figure 6.19 Correlation between wire breaks and relative humidity.

Of course, the true measure of any cable preservation measure is its effectiveness in reducing corrosion and wire breakage. In this regard, a tangible demonstration of the benefits of dehumidification can be seen in the reduction in the number of wire breaks recorded by an acoustic monitoring system installed on a bridge where cable dehumidification was later installed. The correlation between reduction in cable humidity and the reduction in cumulative wire breaks is clearly apparent in Figure 6.19.

Some owners have questioned the feasibility of blowing dry air through the narrow spaces between wires, particularly for cables that have been previously oiled. In such instances, it is recommended that a small-scale blow test be implemented to simulate full-scale conditions. Such a test can be accomplished at a relatively low cost and incorporated into a study phase.

6.5.2.3 Main Cable Preservation: Anchorages

The environment within main cable anchorage interiors is an important consideration in developing an overall cable preservation strategy. These spaces contain the essential elements for transfer of load to the large anchorage block (splay saddles and support columns, eyebar connections, eyebar-to-anchor block interface) and can often be notoriously damp and dirty. A universally accepted preservation strategy is the implementation of an anchorage dehumidification system, consisting of an isolated and sealed enclosure that contains standard dehumidification and control equipment. The isolation enclosure itself can be constructed of a tent fabric or a hardened sheet metal. A hardened structure is recommended when there is the likelihood of falling debris from the anchorage roof (underside of the deck structure).

To maximize the operating efficiency of anchorage dehumidification systems and to generally minimize the corrosive environment within these spaces, anchorage drainage systems should be designed and maintained so as to carry storm water away from the interior space.

6.5.3 Suspender Preservation

Suspender ropes are considered to be replaceable elements; however, due to the high cost of replacement, every effort should be made to reduce deterioration and extend the useful life of these elements. The results of routine and scheduled inspections and sample rope testing will permit the estimation of remaining life, and allow for the programming of intervening improvements and maintenance.

Frequent application of coatings is the most important and effective preservation measure, as is the cleaning and proper maintenance of socket anchorage details. Greater degrees of deterioration are typically found within the socket region, and can be attributed to water that enters and is trapped within the socket, either through "wicking" down the rope or through voids within the rope. Highly confined socket details that are difficult to clean and paint can accelerate the corrosion of rope wire at this troublesome location (Figure 6.20).

Consideration should be given to modifying these details and to providing a means of preventing water from entering by filling any voids in the rope itself with a paste material. At the very least, special attention should be paid to these locations during routine inspections, with priority given to frequent cleaning and painting.

In order to promote and ensure bridge safety, owners should have an understanding of the as-designed and in situ factor of safety of suspender ropes and should establish policies for frequency of inspection and replacement based on these thresholds. An understanding of the interaction of the stiffening truss is also important, in particular the extent to which the truss can carry and redistribute loads safely should a single rope location fail.

In general, when loss of tensile strength is extensive, resulting in a factor of safety of less than 3, replacement of suspender ropes should be considered. While it is feasible to replace the most deteriorated socket region through splicing, the cost of such an approach is often not advisable when considering the remaining life of the parent rope material. This approach may have merit when only a few ropes are in question, or where the age and condition of the parent material are favorable. A more typical strategy is to perform wholesale replacement of suspender ropes. For maximum cost efficiency and constructability, replacement of the entire inventory of suspender ropes is advised and should be combined with necessary modifications to socket anchorage details in order to promote durability and proper long-term maintenance.

Deterioration or wire fraying can also occur at locations where ropes pass through openings in the stiffening truss or through other parts of the structure (such as sidewalks). In these instances, sufficient clearance must be provided between the rope and surrounding structure so as to prevent wearing and deterioration. Steel collars connected to the opening in the structure and clamped to the rope can be effective in preventing relative

Figure 6.20 Deterioration at suspender rope socket anchorage.

movements; however, these details can also create a "dam effect" where water and debris can collect. Frequent maintenance of collars and sacrificial collar liners is important in order to prevent this from occurring.

Suspender rope vibration and oscillation of longer ropes can be an issue of real concern or can serve to unnerve maintenance crews and members of the public. In such instances, spacers or vibration dampers can be used effectively to reduce visible oscillations. As with collars, sacrificial spacer liners should be inspected and replaced when necessary to avoid fraying of wires.

6.5.4 Monitoring Systems

As the ultimate objective of any preservation strategy will be to prolong the life of system elements by slowing the degradation process, monitoring provisions are essential and, wherever possible, should be designed and incorporated into preservation improvements.

6.5.4.1 Structural Monitoring Systems

The purpose of structural monitoring is to provide information on a bridge's response to loadings (temperature, wind, seismic activity, traffic, etc.) compared with the original design intent. Structural monitoring involves the following:

- Installation of various sensors on the bridge, including signal collection and conditioning units
- A data communication system used to transfer data onto a remote computer
- A database application to collect, store, and process sensor data in real time

The design of a structural monitoring system (SMS) needs careful input from the bridge owner to ensure that the information provided by the system is relevant, in a suitable format, simple to operate and maintain, and cost-efficient.

In terms of monitoring the suspension system, any rapid changes of bridge geometry could be a sign of damage (for example, a vehicle collision with suspenders). Longer-term changes in bridge geometry (after correcting for temperature and other loadings), such as sag of the main span, could be an indication of creep of the main cables and hangers or draw of the sockets.

Direct measurements of suspension cable loads are more difficult, as strain gauging of individual wires within the cable would be required, which would involve removal of the wrapping and would probably only involve measurement of outer wires. Load measurement in ropes is also difficult due to the lay of the wires. Accelerometers can also be installed on cables to provide information on vibrations, seismic monitoring, structure modal analysis, and vehicle dynamics.

The bridge owner should be aware that the sensors are generally located in aggressive environments, and their robustness is critical in ensuring a reliable system. Similarly, any wiring should be well protected with suitable armoring. Reliability is a big issue with structural monitoring systems, and the progression of electronic advancement renders some elements obsolete within 3–5 years.

6.5.4.2 Acoustic Monitoring Systems

In order to detect wire breaks within the main cables without carrying out intrusive inspections, acoustic monitoring systems can be installed on the bridge to provide data on the frequency and location of wire breaks along the entire length of the main cables. Electronic sensors are glued onto the cable band castings using structural adhesive between the base of the sensor and the steel casting, typically on every other cable band, either by rope access or from mobile elevated access platforms.

Acoustic monitoring sensors can be hard-wired or wireless. Wireless sensors require an alternative power supply (solar panels, attached to a battery pack, usually mounted onto the main cable hand strand posts). Wired sensors have signal cables attached to the sensor passing down the suspenders, secured to the suspender with stainless steel cable ties, down to the deck to a series of acoustic monitoring units located along the bridge (under the truss or within the deck cross section). These units are connected to a base station via network communication cabling running the length of the bridge. The base station is a server that analyzes data automatically and posts monitoring results via an Internet connection to a remote monitoring application website. Authorized users can access the website at any time to review monitoring data. An automated email or SMS warning system will alert nominated persons of potential wire breaks as they occur (Figure 6.21).

The acoustic monitoring system has a programmed sensitivity (threshold) that is set at the required level to detect wire breaks, but is sufficiently high that background noise emissions are not recorded. Setting the threshold, together with the front-end filters, is a key part of the commissioning process. The system is trained to recognize through signal characterization any detected environmental sounds (such as rain, wind, hail), service sound (such as traffic noise, expansion joint movement), and maintenance

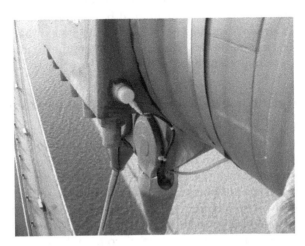

Figure 6.21 Acoustic monitoring device mounted on cable.

activities (such as inspectors or maintenance engineers walking on the cables, painting, etc.). The system typically runs on 110 V main power.

Validation of wire breaks has long been a topic of discussion when reviewing acoustic monitoring systems. It is very difficult to verify fresh wire breaks unless the cable is unwrapped and wedged open to look for evidence, and even then, there are thousands of wires at any one cross section, making detection very difficult. However, acoustic monitoring systems do seem to provide convincing data on the whole, particularly when monitoring systems are used in conjunction with information provided during intrusive inspections. One particular example, on a UK bridge, was that the acoustic monitoring system detected a wire break that occurred overnight (between shifts) in a very localized area of the cable where a short section of the cable had been unwrapped. The wire break was located the next morning, and it could be seen that the ends of the wire were shiny, indicating a recent break.

Experience during intrusive inspections is that the noise created by wedging and inspection works themselves causes the acoustic monitoring systems to struggle to cope with processing all the data. Often, the acoustic monitoring system is paused during work shifts and switched back on between shifts. This causes potential loss of wire break data and is an area for signal filtering development. Acoustic monitoring results have indicated that wires with cracks close to breaking are effectively "tipped over the edge" during intrusive cable inspection work, with increased wire breaks detected during work, but dramatically falling off after completion of the inspection activity.

It is incumbent upon bridge owners to understand the functionality of acoustic monitoring systems and, in particular, how results may be skewed or affected by changes to the bridge or other external events. As pointed out, wire event data should be evaluated against any bridge maintenance or operational activity that may have been occurring that would account for the anomaly. Similarly, the system must be calibrated or "tweaked" to reflect changes or modifications to the suspension system that may impact the attenuation of sound waves. Routine and programmed maintenance of the system is also important in order to ensure the reliability of data.

6.5.5 Operational and Maintenance Considerations

Consideration of operational and maintenance (O&M) impacts must go hand in hand with the evaluation of suspension system preservation strategies. Of particular importance is how new materials and systems can best be incorporated into the owner's existing O&M regime. Key questions revolve around adequacy of existing skill sets, additional training and certifications that may be required, purchasing of new or specialized equipment, and impacts to ongoing (and increasingly constrained) operational budgets. Wherever possible, use of commonly used materials and standard off-the-shelf equipment that can be quickly sourced and procured should be given preference. Where specialized materials or technologies are selected, such as technologies and software associated with structural or acoustic monitoring systems, it is best to utilize an outside contract with specialty vendors (often the same vendors who installed the system) to provide O&M support.

Cable dehumidification systems are designed for minimal maintenance. The wrapping and sealing should be almost maintenance-free for their expected life span of (typically) 30 years, after which it is relatively easy to remove. Mechanical damage to wrapping and sealing can be easily repaired. The plant rooms are normally inspected and maintained twice a year by trained technicians. Routine inspection and maintenance is usually carried out over a 1-day site visit. Usually maintenance involves replacing filters and minor wearing parts. The control system is also checked for correct programming and operation. The wrapping, caulking, sleeves, ducting, and wiring are checked by walking along the main cables at least once a year.

The emergence of hardening and other security measures associated with suspension bridge systems further complicates the consideration of system preservation measures. Special attention must be paid to how these security measures might inhibit or obstruct various preservation measures.

6.5.6 Commentary on Cable Strengthening and Replacement

As previously indicated, preservation of the main cable is of utmost importance, as replacement will often require extensive financial resources and have major impacts on bridge operations. In those instances where preservation alone cannot ensure the long-term safety and viability of the cable, owners must plan for strengthening and eventual replacement. Interim measures such as vehicle weight restrictions may also be required; however, the impact of these restrictions is often far greater than the benefit. Frequent and ongoing monitoring is essential in order to provide sufficient advance notice that will avoid such drastic measures impacting bridge operations. Given the uncertainty of internal wire conditions that can suddenly come to light during internal cable investigations, owners of older bridges with cables approaching threshold safety factors would be wise to put in place contingency plans that can be implemented quickly. Such plans should include measures to repair or augment the main cable system, and studies of the feasibility of cable replacement.

Chapter 7

Maintenance of Suspension Bridges

Cable Dehumidification

Katsuya Ogihara

CONTENTS

7.1 INTRODUCTION

Main cables used in suspension bridges are the most important members of the bridge. It is necessary to keep the main cables in good condition in order to avoid replacing the main cables or even reconstruction of the bridge.

Various in-depth investigations showed that conventional cable protection systems using paste, wrapping wires, and paint coating were insufficient for good corrosion protection of the main cables of a suspension bridge. A cable dehumidification system, sending dry air inside the main cables and eliminating internal water and high humidity, was developed from the results of various tests and investigations. The system was called a dry-air injection system for main cables of suspension bridges in Japan.

The system was introduced for the first time in the world for the Akashi Kaikyo Bridge, one of the Honshu-Shikoku bridges, at the time of its opening in 1998. The system was also installed and airtight works completed in all Honshu-Shikoku suspension bridges by 2002. The system is now also in most of Japan's long-span suspension bridges and in at least 13 suspension bridges in other parts of the world. Several bridges are under installation or under planning right now.

Since the system was developed in Japan, and more than half of the bridges that have the system are in Japan, this chapter explains a Japanese system as a typical example.

The basic specifications are as follows. There are some variances depending on the differing conditions for each bridge:

- The air pressure at the injection point is below or around 3.0 kPa.
- The distance between an injection point and an exhaust point along the main cable is below 200–300 m.

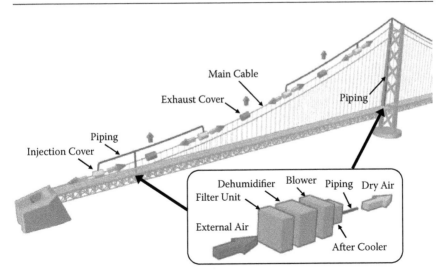

Figure 7.1 Typical cable dehumidification system.

- The airflow volume at the injection point is 0.25–1.5 m³/min, depending on the diameter of the main cable.

The system has been modified with monitoring data and repair results.

A typical dry-air injection system for a main cable consists of a filter unit, a dehumidifier unit, a root blower, an after cooler (optional), piping, air injection covers, and air exhaust covers (Figure 7.1).

There are three kinds of airtight systems for a main cable: a paint coating on the round wrapping wires, a paint coating on S-shaped wrapping wires, and a paint coating with rubber wrapping over round wrapping wires (Figure 7.2). Flexible fluorine resin paint is used on the wrapping wires because of its flexibility.

7.2 CONVENTIONAL CABLE PROTECTION SYSTEM

The conventional main cable protection system for suspension bridges consists of paste on the parallel wires, wrapping wires on the paste, and paint coating on the wrapping wires (Figure 7.3). The purpose of the system is to prevent water intrusion into main cables.

The longer the center span length of a suspension bridge, the higher the share of the weight of main cables in the bridge. The center span length of the Akashi Kaikyo Bridge is the world's longest, and the share of the dead load in the tensile force of the main cables is much higher than that of any prior suspension bridge (Figure 7.4). Therefore, for the Akashi Kaikyo Bridge, the strength of the wires for main cables was upgraded

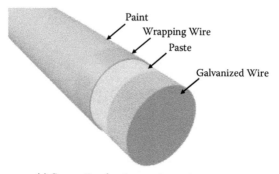

(a) Conventional anti-corrosion system

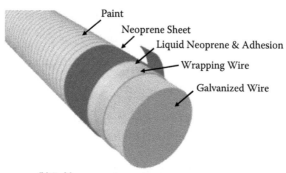

(b) Rubber wrapping system

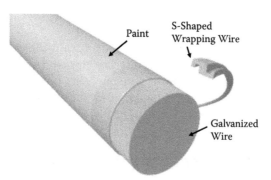

(c) S-Shaped wire wrapping system

Figure 7.2 Airtight systems for main cables.

from 160 kgf/mm^2 (1570 N/mm^2) to 180 kgf/mm^2 (1770 N/mm^2), and the safety factor of the main cables was lowered from 2.5 to 2.2 in order to reduce their weight. Because of this background, it was important for the main cables with higher strength and a lower safety factor to have more effective protection. The study of a more effective cable protection system by the Honshu-Shikoku Bridge Authority (HSBA) started in 1988.

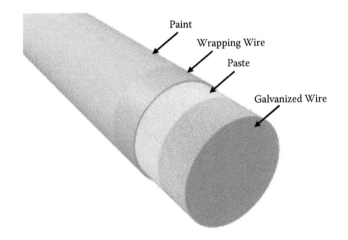

Figure 7.3 Conventional cable protection system.

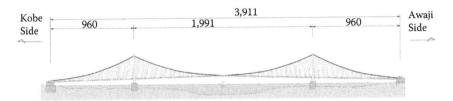

Figure 7.4 Akashi Kaikyo Bridge.

The investigation inside the main cable itself was carried out with unwrapping work and wedging into the main cables of the Innoshima Bridge in 1989. Only 6 years had passed since the opening of the bridge in 1983. Still, the investigation found that some water intruded inside the main cables, and some corrosion was found on some wires (Figure 7.5). Other Honshu-Shikoku suspension bridges that had already been opened to traffic were also investigated between 1991 and 1993. As a result, it was found that the main cables of all bridges were corroded. Corrosion was found not only in the middle of the center span, but also along the entire main cables of some bridges. In addition, it was found that water intruded from paint cracks, and the paste deteriorated, turning the cracked area into a water-retaining body. Since some corrosion was found in the Seto Ohashi suspension bridges just 4 years after their completion, it was conjectured that the corrosion started in the early stages of the service of the bridges.

Nearly all major suspension bridges in Japan, except the Honshu-Shikoku bridges, were also investigated between 1989 and 1995. In the investigation of the main cables of the Wakato Bridge, 27 years after its opening in 1962, although water leakage from the main cables was found, the main cables were in fine condition. Zinc chromate paste, which was used in the bridge

Figure 7.5 Unwrapped main cable of Innoshima Bridge in 1989.

and is no longer allowed to be used, was conjectured to provide high-performance protection.

On the other hand, the inside of the main cables of the Higashi Oi Bridge, 15 years after the opening, had corroded, despite the paste being the same as that used in the Wakato Bridge. The paste of the bridge was fine, and the portion of wires having contact with the paste was well protected.

The HSBA conducted a series of cable model tests with many kinds of paste. As a result of the tests, some high-performance pastes worked well in protecting main cables from corrosion. Conversely, when water intruded inside the main cables, the inside, not the surface, of the main cable corroded even when high-performance paste was used on the cable surface.

7.3 DEVELOPMENT OF A CABLE DEHUMIDIFICATION SYSTEM

7.3.1 Feasibility of a Cable Dehumidification System

From the results of the investigations of the cables of the existing bridges and cable model tests, it was found that over extended periods, the conventional cable protection system could not stop water intrusion into main cables.

Alternatives to the conventional cable protection system, airtight wrapping and dry-air injection, had been studied with cable models and site investigations since 1994 in order to evaluate the feasibility of a dry-air injection system.

The performance of dry-air injection for preventing white-rust production was examined by accelerated tests with a cable model. As a result of the test with 3 months of continuous dry-air injection, although white rust was found at the lower portion of a cable model where water was left for a month before all of it was eliminated by the injection, other portions of the cable model stayed in fine condition. On the other hand, as a result of the same conditions without dry-air injection, white rust was found all over the cable model. By comparing the results of these tests, the performance of a dry-air injection system was verified.

7.3.2 Injection from a Part of a Section of the Main Cable

In order to know the effect of dry airflow inside the main cable, the process of drying out inside the cable model was observed. The model was 2 m long and had a diameter of 60 cm and 11,557 wires with a void ratio of 18%. Water (2500 ml) was poured inside the main cable at the beginning of the test. Dry-airflow volume was 1.0 m³/min. Dry air was sent from a part of the main cable section. As a result of the test, water inside the cable dried out in 300 h. It was found that injection from a part of the section was able to dry out whole sections of the main cable.

In order to know how much water volume could intrude inside the main cable, the volume was measured with the same cable model. As a result, the maximum water intrusion volume came to a void ratio of about 15%.

7.3.3 Injection from the Surface of the Main Cable

Because dry air was expected to be injected from the surface of the main cable, a larger-scale cable model was used to confirm the spread of dry air into the entire section. As a result, 0.19 m³/min dry air could be sent through a section of 13×5 cm on the surface of the main cable using an air pressure of 0.9 kPa. It was conjectured that dry air could be sent from the surface of the main cable and could reach the entire section of the main cable.

7.3.4 Critical Humidity to Start the Corrosion of Galvanized Steel Wires

For steel members to corrode, water and air are necessary. If galvanized steel wires have some substances on the surface, such as rust, paste, or salt, the lowest-humidity level for corrosion occurrence, that is, the critical humidity to start corrosion of galvanized steel wires, was expected to become lower than that in general atmospheric environment conditions. It was important to know the critical humidity at which corrosion of the main cable starts at the site in order to control the dry-air injection system in actual conditions. It was known that galvanized wires do not corrode under a relative humidity of less than 60% in a general atmospheric environment. In order to identify the critical humidity related to the volume of attached substances, several tests were conducted. As a result, not

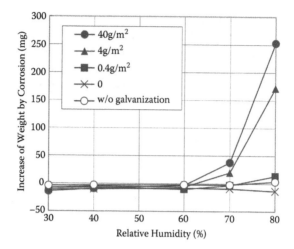

Figure 7.6 Critical humidity to start corrosion of galvanized steel wire.

depending on the volume of attached salt, almost no corrosion was found under the humidity condition of less than 60% RH (Figure 7.6).

7.3.5 Possible Drying-Out Distance

In order to obtain the optimum distance between the injection cover and its exhaust cover, several site experiments were examined. From the experiments, the possible drying-out distance was obtained; the figures were 140 m over 6 months and 215 m over 1 year in the Kita Bisan-Seto Bridge.

7.3.6 Cable Dehumidification System

From these test and experiment results, specifications of a dry-air injection system were determined. A basic system consists of a filter unit, a dehumidifier, a blower, pipes, injection covers, and exhaust covers. The target relative humidity at the injection point was set equal to or below 40%. If some salt is contained in the injected dry air, it accumulates inside the main cable over extended periods of operation, which is not conducive to good main cable conditions. In order to avoid this situation, a filter unit removes salt particles.

7.4 INSTALLATION OF A CABLE DEHUMIDIFICATION SYSTEM

7.4.1 Honshu-Shikoku Bridges

System installations for the Honshu-Shikoku suspension bridges started with the Akashi Kaikyo Bridge. Following the Akashi Kaikyo Bridge, the system was installed in the Kurushima Kaikyo Bridges, and other Honshu-Shikoku suspension bridges in order. Test operations of the system added to existing bridges starting in 1997, in conjunction with installation in newly built bridges. Table 7.1 shows the systems in Japan.

7.4.1.1 Akashi Kaikyo Bridge

Planning and design work for the dry-air injection system into the Akashi Kaikyo Bridge started in its construction stage. The bridge, about 4 km long, was completed in 1998. Waterproof airtight rubber wrapping was applied over the wrapping wire of the main cables. Thirty-two injection covers and eight operation units were installed. Paste was not used because paste was considered capable of becoming a water-retaining material. The separation between injection points and exhaust points was set at about 130 m, considering the drying-out duration of the entire main cables over 1 year.

7.4.1.2 Kurushima Kaikyo Bridges

The Kurushima Kaikyo Bridges comprise three suspension bridges and were completed in 1999. Newly developed S-shaped wrapping wire was applied to the bridges to improve airtightness and waterproof performance.

The First Kurushima Kaikyo Bridge, 960 m long, has eight injection points on its main cables and two operation units. The maximum distance between an injection point and an exhaust point is about 180 m. Twelve injection points and three operation units were installed originally in each the Second Kurushima Kaikyo Bridge, 1515 m long, and the Third Kurushima Kaikyo Bridge, 1570 m long. The maximum distances between an injection point and an exhaust point were about 140 and 160 m, respectively. The system installation was decided during the construction stage, as was done for the Akashi Kaikyo Bridge.

7.4.1.3 Ohnaruto Bridge

The Ohnaruto Bridge, 1629 m long and completed in 1985, has conventional round wrapping wires. Thirteen injection points and two operation units were installed in the bridge between 1997 and 1998. The maximum distance between an injection point and an exhaust point was about 190 m at that time.

7.4.1.4 Seto Ohashi Suspension Bridges

All three Seto Ohashi suspension bridges, completed in 1988, have conventional round wrapping wires. The Shimotsui-Seto Bridge, 1400 m long, originally had six injection points and three operation units. The system was installed in 1999. The maximum distance between an injection point and an exhaust point was about 250 m at that time.

Table 7.1 Cable Dehumidification System in Japan

Bridge Name	Central Span Length (m)	Bridge Length (m)	Year		Location	Number of Injection Points		Number of Units	
			Opened	System Installed		Original	Present	Original	Present
Akashi Kaikyo	1991	3911	1998	1997	Hyogo	32	32	8	6
Ohnaruto	876	1629	1985	1997	Hyogo Tokushima	13	27	2	2
Shimotsui-Seto	940	1400	1988	1999	Okayama Kagawa	6	10 (12)ᵃ	3	3
Kita Bisan-Seto	990	1538	1988	1998	Kagawa	6	12	3	3
Minami Bisan-Seto	1100	1648	1988	1999	Kagawa	6	12	3	3
Innoshima	770	1270	1983	1998	Hiroshima	8	8	2	2
Ohshima	560	840	1988	1998	Ehime	8	8	2	2
First Kurushima Kaikyo	600	960	1999	1999	Ehime	8	8	2	2
Second Kurushima Kaikyo	1020	1515	1999	1999	Ehime	12	8	3	2
Third Kurushima Kaikyo	1030	1570	1999	1999	Ehime	12	8	3	2
Akinada	750	1175	2000	2000	Hiroshima	8	8	3	2
Hakucho	720	1380	1998	2000	Hokkaido	10	10	5	5
Toyoshima	540	825	2008	2008	Hiroshima	8	8	2	2
Hirado	460	665	1977	2008	Nagasaki	8	8	2	2
Rainbow	570	798	2000	UC	Tokyo	Under installation			

ᵃ ( ) is planned.

Six injection points and three operation units were installed originally in the Kita Bisan-Seto Bridge, 1540 m long, between 1998 and 1999, and the Minami Bisan-Seto Bridge, 1650 m long, in 1999. The maximum distances between an injection point and an exhaust point were about 280 and 290 m, respectively.

7.4.1.5 Innoshima Bridge and Ohshima Bridge

The Innoshima Bridge, 1270 m long and completed in 1983, and the Ohshima Bridge, 840 m long and completed in 1988, have conventional round wrapping wires. The system has eight injection points and two units in each bridge. The systems were installed in both bridges in 1998. The maximum distances between an injection point and an exhaust point are about 210 and 150 m, respectively.

7.4.2 Other Bridges in Japan

7.4.2.1 Akinada Bridge and Toyoshima Bridge

The Akinada Bridge, 1175 m long and completed in 2000, has S-shaped wrapping wires. Four injection points were installed for each side's main cables, and two operation units were installed during the construction of the bridge. The maximum distance between an injection point and an exhaust point is about 200 m.

The Toyoshima Bridge, 825 m long and completed in 2008, also has S-shaped wrapping wires. Four injection points for each side of the main cables were installed, and two operation units were also installed during the construction of the bridge. The maximum distance between an injection point and an exhaust point is about 140 m.

7.4.2.2 Hakucho Bridge and Hirado Bridge

The Hakucho Bridge, 1380 m long and completed in 1998, has S-shaped wrapping wires. The system was installed in 2000. The system has 10 injection points and 5 operation units. The maximum distance between an injection point and an exhaust point is about 190 m.

The Hirado Bridge, 665 m long and completed in 1977, has conventional round wrapping wires. The system was installed in 2008. The system has eight injection points and two operation units. The maximum distance between an injection point and an exhaust point is about 120 m.

7.4.3 World Bridges

The system has been installed in at least 27 suspension bridges in different parts of the world. Fourteen bridges are in Japan. The others are the Forth Road Bridge (UK), Humber Bridge (UK), Severn Bridge (UK), High Coast Bridge (Sweden), Alvsborg Bridge (Sweden), Runyang Bridge (China), Taizhou Yangtze River Bridge (China), Yi Sun-sin Bridge (South Korea), Yeongjong Bridge (South Korea), Sorok Bridge (South Korea), Little Belt Bridge (Denmark), Aquitaine Bridge (France), and Hardanger Bridge (Norway).

7.4.4 Bridges under Installation or Planning

The system is being installed in the Chesapeake Bay Bridge (United States), Rainbow Bridge (Japan), Great Belt East Bridge (Denmark), Izmit Bay Bridge (Turkey), and First and Second Bosphorus Bridges (Turkey).

The system installation is planned for the Benjamin Franklin Bridge (United States), Third Bosphorus Bridge (Turkey), and Kanmon Bridge (Japan).

7.5 PRESENT SITUATION OF A CABLE DEHUMIDIFICATION SYSTEM

7.5.1 Operation and Monitoring

7.5.1.1 Operation

The system maintenance activities on all 10 Honshu-Shikoku suspension bridges are as follows:

- Annual review of the system operation: All system operations are reviewed annually based on the annual measurement results and repair records.
- Annual measurement of the system: Air pressure, humidity, temperature, and airflow volume are measured annually at most of the dehumidifier units, piping near root blowers, injection covers, exhaust covers, and some additional points. These measurements are made manually.
- Monitoring: All Honshu-Shikoku suspension bridges have dedicated automatic measurement systems for the dry-air injection system. The system automatically measures humidity and temperature daily at selected points.

Some data can be unreliable due to sensor malfunction. Air leakage from the sealing of cable bands in all bridges is found every year (Figure 7.7).

7.5.1.2 Monitoring Method

In order to keep the relative humidity (RH) inside of the main cable equal to or below the target humidity of 40% RH, monitoring the humidity at the exhaust cover is the best judgment criterion for determining whether

Figure 7.7 Air leakage at cable band.

the system works well. Each bridge has its own automatic monitoring system, and the results are fed back for operation. Some systems monitor not the whole system, but only some portions. In order to control all of the systems of Honshu-Shikoku suspension bridges at the same level, in June 2011, the Honshu-Shikoku Bridge Expressway Co., Ltd. (HSBE), successor of the HSBA, established a manual with measurement standards for dry-air injection systems. Operating conditions of all systems should be observed by annual measurement according to the manual. Evaluations of the system and the formulation of repair plans for all Honshu-Shikoku suspension bridges are made based on the results of these observations.

7.5.1.3 Annual Measurements

Relative humidity fluctuates with the seasons. The humidity inside the main cables of some bridges is often higher in summer. Accordingly, it was decided that annual measurements are best conducted in summer. The measurements are conducted on good weather days. Measurement items include the relative humidity and temperature at the injection points, exhaust points, and blowers; the air pressure at injection points; and visual observation using the monitoring windows of the injection and exhaust covers (Figure 7.8).

7.5.2 Present Conditions

7.5.2.1 Akashi Kaikyo Bridge

The system was installed in the Akashi Kaikyo Bridge during its construction stage. One hundred fifty days after the system operation start, the inside of the main cable was nearly completely dry. The wrapping system

Figure 7.8 Monitoring window of injection cover.

of the bridge has high airtight performance. Air leakage is relatively low. Humidity higher than the target 40% RH was recorded in summer. The system was improved by reusing the extra dry air, which had been exhausted, and has operated favorably since this modification.

Because of the smooth progress of drying out inside the main cable and the system worked favorably, two operation units could be stopped; the number of operating units was reduced from eight to six. Thirty-two injection points have not been changed. Thirty-four exhaust points have also not been changed. Air pressure at the injection cover has been reduced to below about 1.5 kPa because of smooth operation. The size of root blower motors has been made smaller.

To keep the inside of the main cable dry, the airtightness of the main cable is important. Recent investigation results show that the number of air leakage points at the cable bands has been increasing. This increase does not seriously influence the humidity inside the cable. The main cause of air leakage is attributed to deterioration of sealing material.

Deterioration of the paint coating over the rubber wrapping was also studied. The loss of paint film was less than expected. The investigation of paint film thickness was conducted in order to make a repainting plan for the future.

On the other hand, since measurement equipment used for automatic monitoring can experience some malfunctions because of aged deterioration, some sensors have been replaced in order to maintain data accuracy.

7.5.2.2 Ohnaruto Bridge

The environmental conditions of the Ohnaruto Bridge are among the most severe for the Honshu-Shikoku suspension bridges. Test operations of the system for some portion started in 1994, and the entire system started operation in 1999. System monitoring started in 2001.

High-humidity situations at some points could not be changed for a long time. There was water remaining inside the main cables several years after the system installation. Remaining water was drained from a hole opened manually at the lower-side sealing of some cable bands. Injection points were added, and some exhaust points were modified into injection points; the number of injection points is 27, double the number of the original. The maximum distance between an injection point and an exhaust point is about 160 m now.

The system has been also improved by repairing air leakage. With these improvements, the humidity inside the main cable has continued to improve, although humidity at some points still exceeds the target 40% RH. Continuous improvement of airtight measures is planned.

7.5.2.3 Kurushima Kaikyo Bridges

The air pressure of the system of the Kurushima Kaikyo Bridges was set below 3.0 kPa. Measurement results showed that the performance in summer did not satisfy the target humidity of 40% RH. In order to improve the system, performance of the dehumidifier was improved. As a result, the humidity of the bridge is now lower than 20% RH year-round.

Because of the sufficient performance of the system of the second and third bridges, one unit at the center was stopped, and only two units have been operated since 2001. As a result, only four injection points on each main cable of each bridge are operating, and the maximum distance between an injection point and an exhaust point is about 390 m each. The systems of both bridges have improved in their performance of dehumidification, making their operation economical.

7.5.2.4 Other Bridges in Japan

The drying speed inside the main cable of the Shimotsui-Seto Bridge was faster than those of the other Seto Ohashi suspension bridges. Although the performance of the system was the best of the three, it was still insufficient in summer. The maximum distance between an injection point and an exhaust point was about 250 m. There are 10 injection points on the cable now. The number of injection points will be increased to 12 within a couple of years. The maximum distance between an injection point and an exhaust point is to be about 130 m—just half the original—with the condition of 12 injection points being installed.

High-humidity conditions at some points of the Kita Bisan-Seto Bridge and the Minami Bisan-Seto Bridge could not be changed for a long time. Water remained inside the main cables several years after the system installation. Remaining water was drained from a hole opened manually at the bottom sealings of some cable bands. The number of injection points was increased to 12 at the Kita Bisan-Seto Bridge and the Minami Bisan-Seto Bridge. The maximum distance between an injection point and an exhaust point is about 150 m, just half the original.

The systems of the Innoshima Bridge and the Ohshima Bridge have operated smoothly since the installation. There is no change to the numbers of injection points and operation units.

For the Akinada Bridge, the Toyoshima Bridge, the Hakucho Bridge, and the Hirado Bridge, no modifications have been made. The systems have operated smoothly since their installation.

7.5.3 Evaluation

Since the corrosion of each wire in the main cable cannot be observed directly because of the wrapping wires over the main cable, investigation of the wires has proceeded by removing the wrapping wires at some portions. No red rust was found in the bridges that had systems installed before opening for traffic. No progress of deterioration was found in the bridges in which the system was installed several years after their completion, according to the results of the investigation conducted 10 years after the system installation.

7.5.3.1 Akashi Kaikyo Bridge

The wire investigation where the wrapping wires in the middle of the center span were removed in 2008 was conducted in order to evaluate the system performance. The reason why the point was selected is that it had relatively high humidity levels according to the measurement records, and it was assumed that the position tends to retain water, as it is the lowest point in the center span.

The investigation found no water inside the main cable, and the wires were dry. Some white rust was found on the surface of the wires, and red rust, which has been found in other bridges in which the system was installed after their opening, was not found. Namely, the wires of the main cable of the Akashi Kaikyo Bridge were in fine condition. The cable protection system, composed of a round wrapping wire and a rubber wrapping over it, is almost intact.

7.5.3.2 Ohnaruto Bridge

As a result of monitoring, the humidity at some exhaust covers varied due to the influence of climate, especially rain. It was also found that the position of sensors influenced measurement values.

Investigations involving removal of wrapping wires were conducted at some points, such as near the 5A anchorage, because these points frequently showed high humidity. As a result, it was found that the cables were in good condition and there was no water inside. The surface of wires was investigated from the gap between the cable bands where sealings were removed.

7.5.3.3 *Kurushima Kaikyo Bridges*

Since humidity exceeded the target 40% RH in summer, an investigation was conducted at the exhaust covers in the middle of the center spans of the first and third bridges. As a result, some white rust was found on the surface of the main cable, but wires inside the main cable were found to be in good condition.

7.5.4 Modification

7.5.4.1 *Troubles and Countermeasures*

The systems of the Akashi Kaikyo Bridge and the Kurushima Kaikyo Bridges, installed during the construction stage, remain in good condition. Some air leakage from the sealing at the cable bands and the paint film cracking of the main cable has been found because of aged deterioration.

In other bridges, such as the Ohnaruto Bridge or the Seto Ohashi Bridges, where the system was installed over 10 years after completion, air leakage and paint film cracking were found. Air leakage from the paint film cracking was also found. There were some points in which the target humidity of 40% RH was not always satisfied. The reasons why the humidity level was not satisfied are shown below:

- Air leakage from broken sealing at the cable bands, injection covers, and exhaust covers
- Air leakage from paint film cracking of the main cable
- Air leakage from pipe connectors and from cracks of piping
- Operation stops due to equipment malfunctions
- Malfunction of sensors

Repair work and modification of the system against these troubles were conducted to achieve appropriate cable protection by the system.

The countermeasure is divided into two groups, modification of equipment and repair of the wrapping system. Modification of the equipment is to improve the dehumidifier performance, to shorten the distance between covers by increasing the number of injection covers. The other countermeasure, repair of the wrapping system, is for paint deterioration and for sealing malfunctions in order to keep airtightness.

7.5.4.2 Countermeasures for Airtightness

7.5.4.2.1 Painting

It is not necessary for the bridge with rubber wrapping to worry about air leakage from paint film cracking because the rubber wrapping has high airtight performance.

For the bridges without rubber wrapping, round wrapping wires and S-shaped wires with a flexible paint coating are used in order to keep them in an airtight condition. Paint film cracking has been observed because of aged deterioration. It was found on all of the Honshu-Shikoku suspension bridges.

Although the paint film cracking was found on all of the Honshu-Shikoku suspension bridges, there were some portions where the concentration was much higher. Possible reasons are material characteristics, installation condition, aged deterioration, the difference between the strains of painting, and that of wrapping wires.

Intruded water from the paint film cracking was found in the Ohshima Bridge. The countermeasure for the cracking is necessary.

Although there has been no significant difference on airtight performance between three airtight systems since the system installation, a paint coating on the round wrapping wire system is obviously most economical. The timing of repainting of the round wrapping wires should be considered first in the three airtight systems in the near future.

7.5.4.2.2 Sealing at Cable Bands

The sealing at the cable bands cracked by aged deterioration and separated from the cable bands. Accordingly, the air leakage was found at cable bands, and repair of the air leakage has been conducted. Aged deterioration of sealing material is unavoidable.

7.6 CONCLUSION

7.6.1 Installation

Since conventional cable protection systems using paste, wrapping wires, and paint coating are insufficient for good corrosion protection of the main cables of a suspension bridge, a cable dehumidification system, sending dry air inside the main cable, was developed and used worldwide. The system eliminates internal water and high humidity; after that, the system keeps the inside dry.

Although the system has good corrosion protection performance, the system cannot repair the cable itself and cannot recover the original performance of the cable itself. It is important for the main cables of suspension bridges to install and operate the system as early as possible.

7.6.2 Operation

Air leakage and sensor malfunction are popular troubles of the system from the experience of system operation. Broken sealings at the cable bands, injection covers, and exhaust covers are the most popular sources of air leakage. Other air leakage is from paint film cracking, pipe connectors, and cracks of piping. Some data is sometimes not reliable because of sensor malfunction. The life span of the sensors used in the system is relatively short, and sensor malfunctions do happen. Inspection and repair on the airtight systems and sensors are the key for the cable dehumidification system.

7.6.3 Improvement

When a high-humidity situation at some points of the main cable can not be changed for a long time, the performance of the system is not sufficient. It means dry-airflow volume is not enough. There are two methods to increase dry-airflow volume: high-pressure dry air and shortening the distance between an injection point and an exhaust point. Since high pressure causes air leakage from the sealing of cable bands and covers the main cables, shortening the distance between an injection point and an exhaust point is a preferable solution.

When the system has enough performance, economical operation can be applied by downsizing the blower or the dehumidifier.

7.6.4 Conclusion

It is important for favorable system operation to measure the data continuously and to review and modify the system regarding the measured data and repair results.

Chapter 8

Maintenance of Cable Band Bolts in Suspension Bridges

William J. Moreau

CONTENTS

8.1 INTRODUCTION

The structural stability of suspension bridges is completely dependent upon a few critical structural elements: main cable strength, tower strength, adequacy of anchorages and connections, suspender rope (or hanger) strength, and main cable band bolts. This chapter will focus on the last item, main cable band bolts. All the loads from the bridge are transferred to the main cables by the suspender ropes and their connections. The lower connection of the rope to the stiffening truss is generally a shimmed socket or a pinned connection, both of which must be inspected routinely for corrosion or chaffing of the rope.

In older suspension bridges, the upper connection was generally formed by draping the suspender rope over a cable band. The more modern upper connections consist of a pinned connection between ropes and cable band. This pinned connection improves the redundancy should a single rope fail. In either case, the entire suspended load is carried through friction by simply clamping the cable band to the main cable.

While this critical friction connection may seem to be a weak link in the structural system, slippage of loose cable bands has not been a dominant cause of suspension bridge collapse. This does *not* mean the inspection of the main cables of suspension bridges can assume that cable band tension is adequate, just because there is no indication of cable band slippage. It is recommended that bolt tension be spot-checked on a 5- to 10-year cycle and all bolts retensioned on a 20- to 30-year cycle, or shortly after an intrusive main cable inspection that wedges the main cable open for visual inspection and wire sampling activities. The recompacting operation or the rewrapping of the main cable to replace the seizing wire, after an intrusive main cable inspection, may further compact the overall cable diameter, allowing the cable band bolts to lose some of their tension.

8.2 MAINTENANCE ISSUES

While the collapse of the Kutai Kartanegara Bridge (Figure 8.1) was not deemed to be caused by slippage of the cable bands, it is apparent that the suspender rope system failed. Once a single suspender fails, loads will shift to the adjacent ropes, which may lead to the sequential failure of the next rope, and eventually a full progressive collapse.

The biggest problem in maintaining the proper friction between the cable band and the main cable is the fact that the cable band bolts are compressing a large group of round galvanized wires. These groups of round wires can "nest" together over time, or the zinc galvanizing can begin to "flow" under high compressive stress levels that may form when two high points of the zinc on adjacent wires make point contact. Not all galvanized high-strength bridge wires come through fabrication with a smooth surface. If two adjacent wires are compacted and only points of contact are made, these high points will eventually flatten out somewhat and reduce the overall diameter of the

Figure 8.1 Failure of the Kutai Kartanegara suspension bridge in Indonesia, 2011.

compacted cable. Wires crossed during cable-spinning operations can also create points of contact. If enough of these points of contact of the bridge wires flatten out, the cable band bolts can become loose without the nut ever being turned. A common industry practice is to paint a line across each nut and bolt end after confirming the torque on critical bolted connections. Cable band bolts can lose tension despite this line never moving out of alignment.

Checking cable band bolt tension can be a challenging operation. Typically, many layers of paint will have to be removed to expose the threads of the nut and bolt. Then, a center mark or small hole may need to be drilled in the end of the bolt if a physical measurement is to be taken of the bolt length. Measurements are generally taken with an extensometer. Although bolt length measurements have been taken electronically (acoustic method), it has been the author's experience that measurements cannot be repeated reliably with electronic equipment. Measurement accuracy has to be within one- to two-thousandths of an inch to achieve the necessary accuracy required to estimate bolt tension. Measurements are made of the bolt length with the bolt cleaned but undisturbed, with the bolt loose, and finall,y with the bolt properly tensioned. These measurements will allow the inspector to compare bolt tension before and after retensioning.

Most bridges built in the United States in the 1920s and 1930s used a bolt tension of 55,000 to 65,000 lb on a 1½–1¾ in. diameter bolt. This is equivalent to 0.012–0.014 in. stretch during tightening of a 15 in. long and 1¾ in. diameter bolt, as noted in the equation below:

Stretch or strain $= PL / AE$

$$= 55,000 \text{ lb} * (15 \text{ in.}^2 * 29,000,000) \qquad (8.1)$$

$$= 0.012 \text{ in.}$$

where P is the bolt tension (lb), L is the length of the bolt (in.), A is the cross-sectional area of the bolt (in.^{2}), and E = Young's modulus of the bolt material (lb/in.^{2}).

Torque measurements require a careful correlation between the amounts of force required to twist the nut and the amount of axial force or tension induced into the bolt during nut tightening. Any variation in the surface condition of the threads can make this a difficult correlation to establish. Corrosion that may have formed in the annular space around the bolt shank, within the cable band, will also interfere with the correlation. For these reasons, measuring the "stretch" of the bolt makes for a more reliable check on the direct tension of the bolt and the effective level of friction produced by this important connection.

Similarly, using a jack to directly tension the bolt and then tightening the nut to a snug, tight condition before releasing the jack is the most reliable retensioning method for cable band bolts. Many bridges may not have a cable band conducive to using a jack for direct tensioning. The jack requires

Figure 8.2 Mid-Hudson Bridge main cable inspection and rehabilitation. (From Modjeski and Masters, Mid-Hudson Bridge Main Cable Investigation, Phase III, Modjeski and Masters, for New York State Bridge Authority, Harrisburg, PA, 1991.)

a flat surface, around the perimeter of the cable band bolt, to seat itself and allow the hydraulic jack to pull the threaded portion of the bolt against the cable band without interfering with the nut and our ability to turn and tighten it (Figure 8.6). BIACH Technologies makes a device that was used by the New York State (NYS) Bridge Authority to directly tension the cable band bolt and make snug the nut on the Mid-Hudson Bridge (Figure 8.5). This method was used successfully to retighten all the cable band bolts after the extensive and intrusive main cable inspection (Figure 8.2) that wedged open the main cable over its full length in the early 1990s.

8.3 STATE OF THE PRACTICE

The Mid-Hudson Bridge cable bands did not have enough of the casting exposed around the end of the cable band bolts to allow the use of the jack or direct tensioning equipment. Since many of the bolts exhibited threads with moderate corrosion, developing a system using torque to monitor the bolt tension was not thought to be an accurate and reliable method.

Accordingly, new bolts were proposed, slightly smaller in diameter and higher in strength, so that a shouldered washer could be used, which would keep the bolt centered in the casting and provide the necessary surface for the jack to rest and press against during tensioning operations. Test bolts were stretched in the shop with the BIACH jack, and elongations were measured along with the necessary hydraulic pressure to attain 55,000 lb of tension per bolt (Figures 8.3 and 8.4). These readings were then used in the installation sequence to replace all bolts across the span.

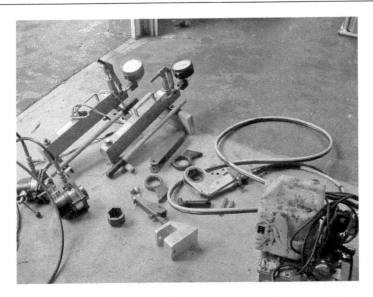

Figure 8.3 Cable band bolt-tensioning equipment. Lower left: Direct tensioning jacks. Above them: Manual hydraulic pump. Center: Hydraulic wrench. Lower right: Power hydraulic pump.

Figure 8.4 Direct tensioning of cable band bolts at the Bear Mountain Bridge using a BIACH jack.

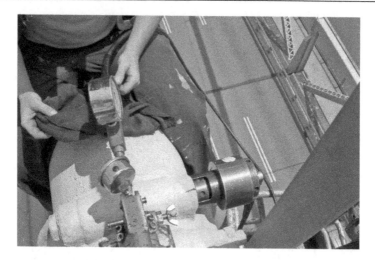

Figure 8.5 Direct tensioning of cable band bolts at the Mid-Hudson Bridge. Note the small holes in the ring of the jack, against the cable band, for snug tightening of the nut after the bolt is tensioned.

Figure 8.6 Required "stick through" of bolt threads and "seating" area on cable band for direct tensioning.

The tension required to achieve 55,000 lb of clamping force is well below the capacity of almost any type of steel bolt. Using a relatively low stress in the bolt provides assurance that the bolt will not stretch over time and create a reduction in the clamping and friction force required for stability of the suspension system.

At the Bear Mountain Bridge (circa 1924), many trials were made with the hydraulic torque wrench (Figures 8.7 and 8.11) and extensometer (Figures 8.8 and 8.9) to establish a well-correlated table of the results of torque to bolt tension. Bolt and nut threads were cleaned to a comparable condition; new bolts were used where old bolts were not reusable. Varying

Figure 8.7 Alternate hydraulic torque wrench where direct tensioning equipment is not feasible.

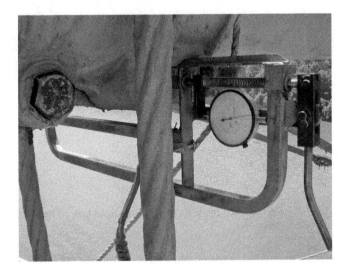

Figure 8.8 Extensometer; used to measure bolt length with a degree of accuracy of 0.002 in. (From Parsons Engineering, Inspection and Condition Evaluation of the Suspension Cables for the Bear Mountain Bridge, Parsons Engineering, for New York State Bridge Authority, New York, 2002.)

Figure 8.9 Field measurement of cable band bolt length using an extensometer.

Figure 8.10 Small hole drilled into the ends of the cable band bolt to improve reliability of extensometer.

torque levels were needed between old and new bolts to establish a uniform bolt tension.

Some of the bolts at the Bear Mountain Bridge could not be replaced without removing the suspender rope, due to interference. The threads on these bolts were thoroughly cleaned and the bolts reused. The first effort

Figure 8.11 Hydraulic wrench tightening of cable band bolt on Bear Mountain Bridge.

to check cable band bolt tension came about after the full-length cable rehabilitation at Bear Mountain Bridge occurred in 2004. A large majority of the bolts were well below the desired tension levels. A spot check was made 5 years later, and less than 10% of the bolts evaluated were below the prescribed tension level. A 10-year cycle was adopted at the NYS Bridge Authority, to spot-check bolt tension levels and perform a full-length retensioning if more than 20% of the bolts are found to be tensioned below the desired level [1, 2].

The Port Authority of New York and New Jersey, as well as the New York State Bridge Authority, attempted to use electronic measuring tools for the bolt length, both in a static condition and during hydraulic torque-tightening operations. No reliability could be established, as the readings would change even when no load was being applied to the bolt–nut assembly. Discrete readings, made only minutes apart, could not reproduce a consistent bolt length. Accordingly, the extensometer (Figure 8.8) was used extensively to establish bolt length. This device can provide reproducible results with a methodical technique used by an experienced technician. The technician will need to practice and develop a routine technique to produce reliable results. A small hole will need to be drilled into each end of the bolt to properly seat the extensometer (Figure 8.10).

Cable band bolts can also be tightened manually. Most people don't have the strength or a wrench big enough to accomplish this with conventional tools. A "torque multiplier" uses gears to increase the torque applied to the nut by a conventional ratchet wrench (Figure 8.12). This method also uses the extensometer to spot-check the stretch of each bolt as it is tightened.

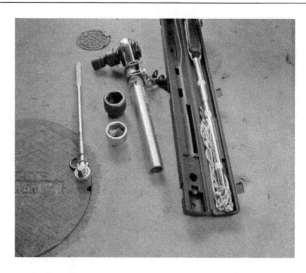

Figure 8.12 Manual torque wrench with a 4:1 torque converter is another option for tensioning bolts.

Careful attention must be paid when choosing to replace cable band bolts and nuts. At the Mid-Hudson Bridge, where a higher-strength replacement bolt was used to reduce the bolt diameter and allow for the BIACH jack to sit properly against the cable band, a locknut was used to secure the principal nut. High-strength bolts are normally tightened to the point where the threads are slightly damaged. This "locks" the nut in place, as vibration can no longer allow the nut to loosen against the damaged threads.

Suspension bridges in Europe, built in the 1960s, used longer spans between suspender ropes and larger cable bands to support larger loads. These more modern bridges tend to use a smaller number of larger and higher-strength cable band bolts. Some bridges in Europe and Asia use a cable band split horizontally, using vertical bolts to clamp them together. In this configuration, the cable band bolts directly support the suspender rope loads, as well as provide the clamping force necessary to transmit the suspended loads through friction. After an attempt at retensioning the cable band bolts, one European bridge owner decided to replace the bolts with higher-strength steel hardware and suffered an unfortunate consequence. During a routine inspection, after only 7 years of service, many of the new nuts were found to exhibit cracking (Figure 8.13). Theories behind the cracking have been attributed to the use of higher-strength steel and less contact area between the nuts and cable band. Damage to the corners of the nuts appears to be further influenced by hydrogen embrittlement concentrated at the high point loading of the corners of the nuts against the cable band. The nuts were replaced using a larger, heavy hex nut made from milder steel without further incident.

Figure 8.13 Cracking of a new high-strength steel cable band nut that recently replaced 35-year-old hardware.

8.4 SUMMARY

The entire weight of the suspended span load is transferred from the vertical suspender ropes or hangers to the main cable simply through friction at the cable bands. Maintaining a static condition of the cable band bolts may not ensure adequate clamping force is provided in this critical connection. This author recommends spot checking of cable band bolt tension on a 5- to 10-year cycle and a more extensive retightening effort every 20–25 years, or after any intrusive main cable inspection.

REFERENCES

1. Modjeski and Masters. 1991. Mid-Hudson Bridge main cable investigation, Phase III. Harrisburg, PA: Modjeski and Masters, for New York State Bridge Authority.
2. Steinman, Boynton, Gronquist, & Birdsall. 1997. Bear Mountain Bridge cable band bolt survey report. Steinman, Boynton, Gronquist, & Birdsall, for New York State Bridge Authority.
3. Parsons Engineering. 2002. Inspection and condition evaluation of the suspension cables for the Bear Mountain Bridge. New York: Parsons Engineering, for New York State Bridge Authority.

Chapter 9

Maintaining Anchorage Enhancements

Russell Holcomb

CONTENTS

9.1 INTRODUCTION

An enhancement is an improvement to an existing system or object. Over time, advances in technology can turn an enhancement into an industry standard, making it difficult to separate anchorage enhancement maintenance from anchorage maintenance. By looking at the experiences of bridge owners in dealing with problems common to the anchorages of suspension bridges of varying ages, span lengths, and service environments, we hope to identify trends and provide general guidance for other owners looking for effective and cost-efficient ways to preserve their bridges. General guidance is the best we can hope for, given the unique character of each anchorage, advances in technology, the small sample size, and bridge owners' justifiable concerns about disclosing contractual or security-related information. Questions (see the appendix at the end of the chapter) were prepared and distributed to bridge owners. We dealt directly with bridge owners as much as possible to minimize the chances of consultant/contractor conflicts of interest, be they actual or perceived. The responses were evaluated to identify common problems, mitigation attempts, and their effectiveness.

9.2 APPROACH

In order to collect information without disclosing sensitive information, questions were broadly written. Not every question applies to every anchorage, and some owners chose not to answer certain questions. Consequently, the number of replies to each question varies. For convenience, answers were divided into three groups named Most, Many, and Some. *Most* means between 51% and 100% of the respondents gave similar answers. *Many* means between 26% and 50% of the answers were similar. *Some* means no more than 25% of respondents gave similar answers. The terms *Many*, *Most*, and *Some* indicate relative values, not actual numbers. To maintain confidentiality, participating bridges are not identified by name in Section 9.4.

9.3 CHALLENGE

For every anchorage, the designer's primary responsibility is to ensure the safe and permanent transfer of load from the bridge to the earth. Once the structural requirements have been satisfied, operational, architectural, and other concerns, including financial constraints, can be addressed and a detailed design produced. However, no two sites, even those at opposite ends of the same bridge, are identical, and aesthetic sensibilities change over time. As a result, each anchorage is, to some degree, unique. Identifying standard maintenance practices becomes more complicated when we consider the age differences among operational suspension bridges. The Brooklyn Bridge (Figure 9.1) was completed in 1883. The Great Belt East Bridge (Figure 9.2) was completed in 1998.

Even the casual observer can see the results of changes to design codes and advances in technology, materials, construction methods, and equipment. Compliance with environmental regulations can also affect the design, operation, and maintenance of a structure. Preventing unauthorized access, always a concern, has taken on extra urgency in recent years, directly impacting maintenance activities. Trying to develop a universal anchorage maintenance plan is clearly impractical. As a result, bridge owners must inspect their anchorages to identify defects with the potential to reduce the capacity of the anchors and develop solutions that will halt the progression of the defect and, if necessary, make repairs. The most complicated repairs could involve reanchoring a cable (Mayrbaurl 1995) or arresting corrosion through a combination of structural alterations

Figure 9.1 Brooklyn Bridge.

Figure 9.2 Storebaelt Great Belt East Bridge. (Courtesy of Kim Agersø Nielsen.)

and dehumidification (Sansone et al. 2006). Both actions are technically demanding and costly. The good news is that reanchoring and reconstruction can be avoided by committing the resources necessary to inspect, protect, and repair the elements of the anchorage on a regular basis. The bad news is that commitments can fade over time. Bridge owners must take that into account and formulate maintenance strategies that use automated systems in addition to traditional methods.

9.4 FINDINGS

This section briefly describes the findings of the survey in five categories: anchorage construction, water penetration, access for inspection, dehumidification, and future development.

9.4.1 Anchorage Construction

Most anchorages in the study are constructed of cast-in-place concrete. Some are built of masonry or a combination of masonry and cast-in-place concrete. One is a reinforced grillage within a steel box girder. Most of the anchorages do not have a secondary function; many do. Of the anchorages with a second function, such as an electrical room or a storeroom, most were originally designed to accommodate those functions. While many anchorages have internal communication systems, most do not. Most anchorages do not have fire alarms, but some do. One has a methane detection system. Animal intrusion was not a problem at most anchorages. Many, however, did have problems with birds and bats. Installing, and maintaining, screens, nets, or other barriers eliminated the problem. Some of the anchorages had drip pans to collect oil dripping from the cable. Most did not. This could be because not all cables are regularly oiled and a few cables are dehumidified. For all anchorages, the locations and numbers of light switches and electrical outlets were considered adequate for routine inspection and repair work. One bridge owner plans to install air compressors in an anchorage to power the repair equipment used on the bridge. Most respondents consider their power supply to be reliable. Some have backup generators.

9.4.2 Water Penetration

Except for one location, water penetration was evident at every anchorage. In most locations, the source was groundwater or river/seawater. Water in the form of rainwater or runoff also entered the anchorages of many bridges. Pumps were installed inside and, when necessary, outside the anchorage to remove groundwater and river/seawater. Rainwater and runoff conditions were addressed by redesign, repair, and cleaning of bridge elements.

9.4.3 Access for Inspection

Many anchorages have permanent platforms for inspection and repair access. Some anchorages rely on temporary platforms. An equal number use a combination of permanent and temporary platforms. Many do not require any access platforms. Platforms were reported to be designed to the appropriate standard. Most anchorages do not have permanent hoists/gantries. Some anchorages either had or will soon have a permanent hoist/gantry. The electrical system of every anchorage was considered adequate for routine inspection and repair work. Most respondents stated that the question about the ease of removing and replacing protective enclosures for the strands and anchors was not applicable to their anchorages. However, one owner reported that the cable-supported plastic fabric that was used to create a dehumidified space was easy to remove and replace. At another anchorage, a metal roof above the cable shields it from water and debris.

9.4.4 Dehumidification

Where the original bridge design included an anchorage dehumidification system, the cost of the system was significantly less than 0.1% of the project cost. All dehumidification systems (Figure 9.3), both new installations and retrofits, were built according to plan, with no significant changes.

Most of the anchorages in the study were dehumidified long after they were constructed. The longest time was 72 years, and the shortest was 10 years. Only three anchorages in the study were originally built with dehumidification equipment. Two have been operating satisfactorily for 44 and 34 years. The third has been in operation for less than a year. Most owners reported no unanticipated problems when installing new equipment in existing anchorages. However, the power supply had to be upgraded at one location, and space constraints had to be overcome at another.

None of the retrofit anchorages were mechanically ventilated prior to installing dehumidification equipment. Regardless of being installed as original equipment or as retrofits, the dehumidification equipment at most anchorages has been replaced, usually in kind. Because our sample is small and the interior of each anchorage is a unique environment and each owner approaches humidity control in a different manner, we cannot state that the equipment at each location will have the same service life. However, we can state that none of the respondents, except one, reported replacing any

Figure 9.3 Dehumidification system at anchorage of a suspension bridge. (Courtesy of Sreenivas Alampalli.)

equipment sooner than the 15–20 years that can be reasonably expected for properly sized and maintained equipment. The exception is at a single bridge anchorage, call it anchorage A, where performance problems became apparent shortly after the system went into operation. The planned installation of higher-capacity equipment is expected to resolve the problem. Diligent owners will not overlook any opportunities to improve performance. At one bridge, when it came time to replace the dehumidification equipment, the owners installed ductwork and moved the unit from inside the dehumidified space to a protected location nearby, making it much more convenient to access the equipment for inspection and maintenance. Removing the equipment from the chambers also contributes to the atmospheric stability of the dehumidified space, as fewer people will have to go in and out of it. In a trial at another bridge, individual cable enclosures, which had been successfully dehumidified for several years, were removed and the entire space dehumidified to provide better access for inspection and maintenance. It was found to be uneconomical to dehumidify the entire space without sealing all the concrete surfaces. The trial ended and the individual enclosures with smaller dehumidifiers were restored. At most of the retrofit locations, corrosion was observed before the equipment was installed. At some retrofit locations, corrosion was not observed, but as stated earlier, the terms *many*, *most*, and *some* are relative.

The following findings apply to retrofits and original installations as well. Except for anchorage A, no active corrosion has been observed in any of the bridges where dehumidification equipment was in operation. Only one owner described the corrosion as "somewhat addressed." At every anchorage, a relative humidity was specified. Every anchorage, except anchorage A, reached the specified relative humidity in the anticipated time frame, and every anchorage, except A, consistently maintained the specified relative humidity. At most locations, relative humidity readings are displayed at the anchorage. At a few locations, relative humidity readings are displayed either at the anchorage and a remote location or just at a remote location. At most anchorages, air temperature is also monitored. Some anchorages do not monitor any other air properties. One location monitors for the presence of methane gas, which is known to exist in the area. The number of monitoring systems that send alerts if the relative humidity goes above a predetermined level and those that do not are almost equal. Most dehumidification systems are maintained by a combination of in-house and contractor personnel. Many are fully maintained by in-house forces, albeit with additional training. The dehumidification system at one bridge is maintained exclusively through a service contract. The maintenance schedule recommended by the dehumidification system manufacturers was considered satisfactory at every location except anchorage A. Traditional cable maintenance activities, like painting or oiling, and the frequency of such activities did not change for most of the retrofitted anchorages. Two locations reported that strands no

Figure 9.4 Protective coating at the eyebar–concrete interface.

longer need to be painted or oiled. However, eyebars at these locations are still greased on a regular basis (Figure 9.4).

Not everyone identified desirable system upgrades, but most of those who did were interested in remote monitoring. Some were interested in monitoring conditions in addition to relative humidity and air temperature. One owner is considering monitoring the relative humidity and temperature at multiple locations to ensure air circulation, monitoring the temperature of the strand anchor rods to observe potential for local condensation, and installing sacrificial corrosion test strips in the anchorage to record long-term trends. Instruments to detect the presence of standing water were also mentioned. Some owners were interested in uninterruptable power supplies.

9.4.5 Future Developments

New paints and protective coatings, including hydrogen-resistant coatings, an effective method of monitoring the condition of eyebars and anchor rods embedded in concrete, impervious paving materials for decks over anchorages, and better concrete sealants were identified as areas warranting further development.

9.5 CONCLUSION

Every bridge owner knows the challenge of maintaining a suspension bridge anchorage. An essential area, generally off limits to all but a few, it must compete with other essential but far more visible bridge components for the funds and services needed to ensure the safety of the traveling public and preserve the bridge. In the United States, the development of National Bridge Inspection Standards required bridge owners to inspect and document bridge conditions in a systematic manner, greatly facilitating the prioritization of work. Regularly updated inspection documents also identify chronic conditions that, if left unattended, would jeopardize the integrity of the bridge. And chronic conditions are, unfortunately, common in anchorages (Hopwood 1981). Interiors can be humid, and water constantly enters the anchorage through cracks in the floors, walls, and ceilings, through worn-out deck joints and drains, poorly sealed cable and conduit penetrations, broken entry hatches, windows, or ventilation louvers, and collects at low points (see Figure 9.5 for a typical finger joint). Closely spaced bars anchored in concrete are difficult to inspect, let alone repair. Anchorages can also become sites of vandalism and antisocial behavior, nuisances at best and fire hazards at worst.

The first step in any maintenance program is getting safe access to the structure, preferably very close to it for thorough visual inspection. Despite advances in aerial work platform technology, inside an older anchorage, access relies heavily on scaffolding and rigging. While a contractor would not be expected to object to working from the anchorage's permanent scaffolding, provided it is in good condition, for legal reasons, a contractor may be reluctant to use temporary scaffolding or rigging erected by others. Additional permanent scaffolding, designed to applicable standards, would minimize the inevitable costs of mobilization and demobilization for every

Figure 9.5 Finger joint drains require regular maintenance to effectively control water.

inspection cycle or corrective action. Similarly, expanding or upgrading the lighting, material handling, and electrical systems would increase efficacy and reduce cost. There are other obstacles to gaining access, namely, traffic and physical security. No bridge owner wants to interfere with traffic, especially on a tolled facility. Some locations require that traffic lanes or sidewalks be closed in order to open a door or hatch, and remain closed to accommodate vehicles and equipment. If lane closures are creating serious problems, owners may want to consider modifying or relocating the entrance. Unauthorized access has always been a concern, and in recent years, the use of sophisticated security systems has become widespread. Many owners are familiar with the restrictions they face when working adjacent to a railroad. Work is approved beforehand, everyone knows their role in the operation, time is strictly limited, and any deviation from the plan can have serious consequences. Performing maintenance in a secured anchorage must be approached in a similar way. Individual workers might need the approval of an authorizing agency before entering the work site. Bridge staff must know what work is planned, who will perform it, when it will start and when it will end, where equipment and materials can be stored, and so forth. Workers and bridge staff must have a reliable means of communication in order to avoid potentially costly work interruptions.

Historically, it was customary for anchorage maintenance programs to have inactive periods between inspections and active periods of repairing conditions found during inspection or responding to emergencies. Preventive measures were limited to the application of protective coatings, although one could argue that maintaining anchorage sump pumps is also preventive maintenance. The introduction of dehumidification changed the maintenance paradigm by effectively eliminating the inactive periods and offering a practical alternative to the protective coatings that were widely used but never completely successful. Dehumidification is a relatively simple technology, and the equipment can be incorporated into supervisory control and data acquisition (SCADA) systems used on large bridges. A SCADA system's ability to remotely control and monitor the conditions in the anchorage and troubleshoot the equipment in the event of a problem can reduce the cost of maintenance by increasing the efficiency of the operation.

In areas where the availability or cost of land precludes expansion, dry anchorages offer a no-cost alternative for the location of telecommunication and other electronic equipment. However, we cannot overestimate the importance of the design process, augmented by feedback from maintenance personnel, when it comes to dehumidification equipment. In our small study, only one anchorage has experienced performance problems. It is reasonable to assume an equally small number of bridges outside the study had or will have comparable problems. In our opinion, this should not discourage others from considering dehumidification.

Of course, not all maintenance is preventive. An anchorage can be a busy place if it has a secondary function. Anchorages of older bridges seem to lend themselves to other uses, ranging from convenient material storage rooms to stages for cultural events (Glureck 1983), putting unanticipated demands on existing utilities or creating a demand for new services. However, water—in the air and entering through cracks in the walls, the roadway/roof slab, defective joints, vents, and drains or through improperly sealed penetrations—is by far the most serious threat to an anchorage. Basic mitigation measures can include sealing cracks in walls and floors, repairing and cleaning existing drainage systems, covering strands with thick plastic shrouds or metal panels to protect them from water and debris, and bringing in pumps several times a year to remove standing water. Solutions that are more permanent may require reconstructing the roadway/roof slabs with dense pavements and waterproof membranes or redesigning and relocating roadway joints to prevent water from leaking directly onto strands, shoes, and eyebars. At the Williamsburg Bridge (Patel et al. 2000), splay castings were moved up the cables, away from the face of the anchorage (Figure 9.6). The cable penetrations were redesigned to prevent rainwater on the cables from entering the anchorage, and a passive dehumidification system, designed to provide continuous airflow throughout the chambers, was installed.

A contract is in place to dehumidify the anchorages of the Brooklyn Bridge (Ahmed 2009), 131 years after they were built. As time goes by, we can expect more suspension bridge owners to enhance their anchorages with dehumidification systems.

Figure 9.6 Redesigned Williamsburg Bridge cable penetrations.

In the United States, tragic events in the last half of the twentieth century made the public aware of the need to improve the bridge inventory. Funds were made available, and many bridges were replaced or rebuilt, often with inconvenience to the community. Bridges represent a considerable investment that should not be squandered through complacency or neglect. The means to preserve them are there. All that is needed is the commitment.

ACKNOWLEDGMENTS

The author sincerely appreciates the contributions of the following individuals and their staff: Barry Colford (Forth Road Bridge), Bill Moreau (Mid-Hudson and Bear Mountain Bridges), Ashok Patel (Benjamin Franklin Bridge), Noel Stampfli (Golden Gate Bridge), Peter Hill (Humber Bridge), Jens Kristian Tuxen (Little Belt Bridge), Kim Nowack (Mackinac Bridge), Tony Hudson (M48 Severn Bridge), Andrew Gordon (and Marwan Nader, T.Y. Lin International) (San Francisco Oakland Bay Bridge), Finn Bormund and Kim Agersø Nielsen (Great Belt East Bridge), and Dave Riggs, Aimen Youssef, and Chris Saladino (Bronx-Whitestone and RFK Triborough Bridges).

REFERENCES

Ahmed, H. 2009. Rehabilitation of the Brooklyn Bridge—Contract 6. MENY.
Glureck, G. 1983. Brooklyn Bridge unveils its own gallery. *New York Times*, May 27.
Hopwood, T. 1981. *Ohio River Suspension Bridges: An Inspection Report.* Research Report UKTPR-81-6.
Mayrbaurl, R.M. 1995. Cable anchorage repairs on New York City bridges. Presented at IABSE Symposium on Extending the Lifespan of Structures, San Francisco, August.
Patel, J.A., Barbas, J.A., and Bruschi, M.G. 2000. Reconstructing the Williamsburg Bridge. Presented at the 2nd ISBOC.
Sansone, R., Jenal, J.R., Minas, A., and Sloan, S. 2006. Dehumidification of George Washington Bridge anchorage chambers. Presented at 5th International Cable Supported Bridge Operators' Conference, August 28–29.

APPENDIX: QUESTIONS TO BRIDGE OWNERS

Bridge name: _____

Part I: Dehumidification

Case A: Original Bridge Construction Contract includes Anchorage Dehumidification

1. What was the cost of the dehumidification system as a percentage of the project cost?
2. Does the as-built system differ from the plan? If so, what was the nature of the change and how did it affect the cost?
3. How long has the system been in service?
4. Has corrosion been observed since the system became operational?
5. Was the system designed to provide a specific relative humidity?
6. Was the specified relative humidity reached in the expected time?
7. Is the specified relative humidity level consistently met?
8. Is the relative humidity level displayed at the anchorage or at a remote location?
9. What other air properties does the system monitor?
10. Does the system send an alert in the event that readings exceed target values?
11. Which maintenance tasks can your personnel perform and which require specialists?
12. Is the manufacturers' recommended maintenance schedule sufficient to keep the system in operation?
13. If funds were available, what upgrades or changes would you make to the system?

Case B: Retrofit Anchorage Dehumidification System to an Older Bridge

1. How many years elapsed between the construction of the bridge and the installation of the first dehumidification system?
2. Were there unanticipated difficulties installing the equipment due to its size, weight, utility requirements, etc.?
3. Did the as-built system differ from the plan? If so, what was the nature of the change and how did it affect the cost?
4. Has the system been replaced or upgraded since it was first installed?
5. Was the anchorage mechanically ventilated before dehumidification equipment was installed?
6. Was the degree of corrosion in the anchorage documented before the system was installed?
7. Has all corrosion stopped since the system became operational?

8. Was the system designed to provide a specific relative humidity?
9. Was the specified relative humidity reached in the expected time?
10. Is the specified relative humidity level consistently met?
11. Is the relative humidity level displayed at the anchorage or at a remote location?
12. What other air properties does the system monitor?
13. Does the system send an alert in the event that readings exceed target values?
14. Which maintenance tasks can your personnel perform and which require specialists?
15. Is the manufacturers' recommended maintenance schedule sufficient to keep the system in operation?
16. Has the schedule for traditional protective work like painting, cleaning, and oiling changed since the system went online? If so, quantify the change.
17. If funds were available, what upgrades or changes would you make to the system?

The following questions apply to *all bridges.*

Part II: Water Control

1. Was, or is, there evidence of water, including groundwater or saltwater, penetrating the anchorage?
2. What was, or is, the source of the water?
3. How was, or is, the water controlled?

Part III: Access for Inspection and Repair

1. Are all the strands and anchors accessed by temporary scaffolds, permanent platforms, or a combination of both?
2. Are the scaffolds and permanent platforms designed to support industry standard workloads?
3. Are there built-in equipment hoists or gantries?
4. Can routine inspections and repairs be performed using the existing electrical system only?
5. Where individual cables are enclosed for dehumidification, dust protection, etc., can sections of the enclosure be easily removed and replaced if additional access is required?

Part IV: General

1. What is the anchorage's primary construction material: masonry, cast-in-place concrete, other?

2. Is anchorage space also used for equipment rooms, workshops, general storage, etc.? If so, was that arrangement part of the original anchorage design?
3. Is there a fire alarm in the anchorage?
4. Is there a communication system to connect the anchorage chambers to each other and to the outside?
5. Have birds or other animals been found in the anchorage? If so, have attempts to keep them out been successful?
6. Is there a permanent system to collect oil that drips from the cable?
7. Are light switches in prominent locations?
8. Are there a sufficient number of electrical outlets and are they conveniently located?
9. Are uninterruptable power supplies consistently reliable? If not, why?

Part V: Future Developments

1. Which technologies, materials, or equipment designed to preserve the anchorage warrant further development?

Chapter 10

Maintenance of Suspension Bridge Anchorages

Stewart Sloan, Danny K. Cobourne, Morys Guzman, and Judson Wible

CONTENTS

10.1 INTRODUCTION

Suspension bridge anchorage maintenance is a critical part of the overall upkeep of a suspension bridge. While not as challenging as suspender rope or deck replacements, the necessity is equal to that of those projects. The George Washington Bridge (GWB) anchorages have withstood the tests of time due to operational maintenance by the Port Authority of New York/New Jersey (PANYNJ) (Figure 10.1). Measures such as regular inspection every 2 years, along with more frequent informal maintenance inspections of critical systems such as the cable dehumidification and flood prevention systems, have provided a stable foundation for the bridge. The anchorage maintenance that is discussed in this chapter focuses on the methods and solutions for the George Washington Bridge.

The New Jersey anchorage was built into rock with 200,000 cubic yards of material removed,[1] and it has faced the majority of bridge maintenance challenges over the years. Major maintenance challenges have arisen by groundwater entering the anchorage, with flood prevention and dehumidification utilized as mitigation measures. This chapter highlights the dewatering and dehumidification methods used to control the water infiltration and humidity levels within the chambers.

Figure 10.1 A view of the George Washington Bridge from the New York side. Photo credit: Port Authority of NY/NJ.

On the other side of the river, the New York anchorage was built above ground with 110,000 cubic yards of concrete. The maintenance challenges that have been successfully mitigated are discussed. Even though the content of this chapter is solely based on the experience of maintaining George Washington Bridge anchorages, similar maintenance strategies apply to most suspension bridge anchorages.

10.2 BACKGROUND

Othmar Ammann, a Swiss-born architect and engineer, proposed a bridge between New York and New Jersey in 1923 that ultimately was chosen from several others to be the Hudson River Bridge, now known as the George Washington Bridge (GWB).[1] The location was chosen due to a narrowing of the river at that location. The port authority held groundbreaking ceremonies on September 21, 1927, with construction beginning shortly thereafter. The towers and floor steel were completed in January 1931.[1] The GWB was first opened to traffic in 1931, with Governor Franklin D. Roosevelt (later President Franklin D. Roosevelt) at the dedication ceremony. It was completed 6 months ahead of schedule for less than the original $60 million budget. Five-and-a-half-million vehicles used the original six-lane roadway in its first full year of operation in 1932. At the time it was constructed, it was the longest suspension bridge in the world, with a main span of 3500 ft and balanced spans on each side, as shown in Figure 10.2. At the time, it had nearly twice as long a main span as any other bridge in the world. The original design as built contained a significant factor of safety to take into account possible additional future lanes or modes of transportation.

10.3 BUILDING THE ANCHORAGES

With the bottom of the pit roughly 130 ft belowground, building the two New Jersey anchorage chambers provided considerable construction challenges. As shown in Figure 10.3, the two chambers are a considerable distance apart, which provided a benefit for the contractor during excavation to allow shifting work between the two chambers during blasting. Figure 10.4 shows the progression of excavation and removal along with the quantities. Nearly 200,000 cubic yards of material was excavated for the New Jersey anchorage of the bridge, which anchors the main cables for the bridge on the New Jersey side of the river. Most of that material was stone from the 300 ft high Fort Lee granite cliffs, which are located roughly 500 ft away from the shoreline of the Hudson River.[1]

When building the New Jersey anchorage, mats of reinforcing bars were provided on the faces of the embedding concrete, guarding against concrete shrinkage and preventing groundwater from reaching the anchorage steel. In addition, construction joints were avoided by continuously pouring the concrete. During construction, the anchorages experienced water infiltration, and sump pumps equipped with automatic float switches were used to pump the water from the excavation. The complex and challenging anchorage excavation and concrete block work was required to be completed by October 1, 1928, so that wire-spinning operations on the main cables could proceed. The rest of the cut outside the anchorage was completed on January 8, 1931.[1]

The New York anchorage consists of 110,000 cubic yards of concrete, which was completed on March 13, 1929. The Washington Heights ridge rose 200 ft high only 1000 ft away from shore. The massive concrete block structure built for the anchorage weighs 220,000 tons and anchors the main cables. Sited plan dimensions of the anchorage are roughly 300 ft longitudinally and 200 ft perpendicular to the bridge, and when constructed, it was shaped as a U, as shown in Figure 10.5. The large amount of concrete necessitated a concrete plant on site with materials brought in barge.

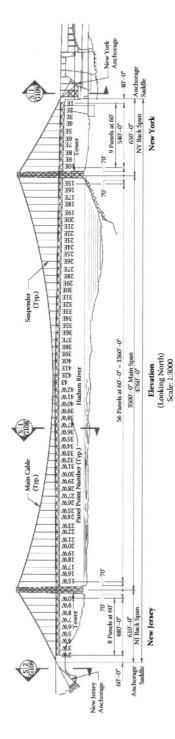

Figure 10.2 Elevation of the bridge looking north. Note the short back spans. Photo credit: Port Authority of NY/NJ.

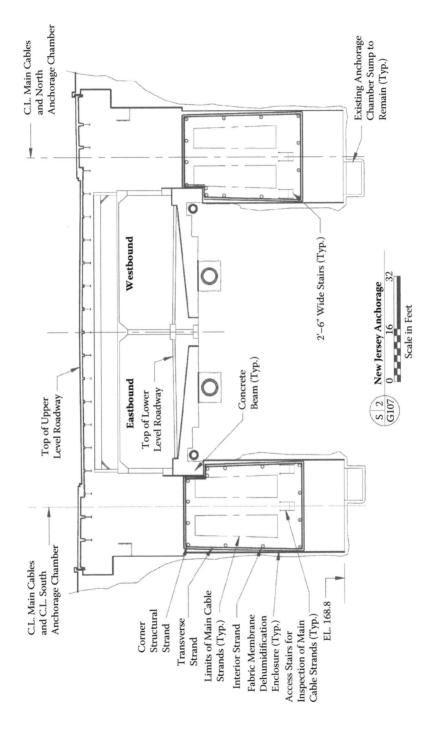

Figure 10.3 Cross section of the New Jersey anchorage and dehumidification measures. Photo credit: Port Authority of NY/NJ.

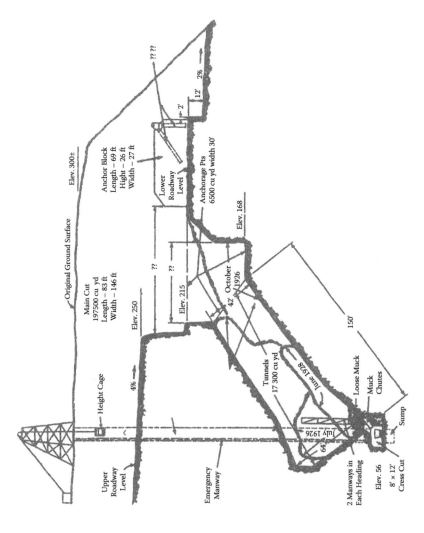

Figure 10.4 A view of the excavation method and quantities for the New Jersey anchorage. (With permission from ASCE.)

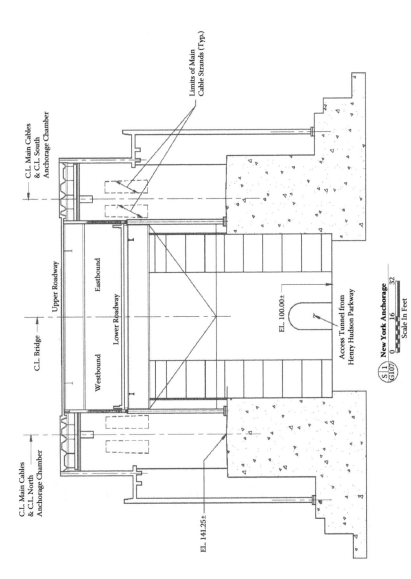

Figure 10.5 Cross section of the New York anchorage. Photo credit: Port Authority of NY/NJ.

10.4 THE NEW JERSEY ANCHORAGE

Due to their construction underground, each of the two New Jersey anchorage chambers are subject to high hydraulic gradient, causing the percolation of groundwater through the walls of the chambers. Groundwater is collected from the walls by a passive drainage system of PVC gutter troughs that are placed at different locations abutting the anchorage walls, and gravity fed to a sump pit at the bottom of each chamber. Each sump pit is approximately 5 ft 6 in. wide by 8 ft 6 in. long and 4 ft 0 in. deep; the pit and pump configuration can be seen in Figure 10.6. The pits have reinforced concrete walls. The main drainage system of PVC troughs and piping is supplemented with additional rectangular drainage sections built into the floor in certain locations, providing passive drainage to the sump pits. Both chambers have steady flow rates, with the higher flow occurring in the southern chamber.

The southern chamber of the New Jersey anchorage has had several flooding events in the past. On January 21, 2003, a flooding event occurred after the pumps were blocked by debris that entered the south chamber sump pit. The estimated volume of water infiltration for that event was 125,000 gal, or over 1 million lb of water, which covered several of the cable anchorages. Portable sump pumps were brought in to remove the water. At the time, there were two sump pumps within each chamber of the anchorage. Subsequent to that, a third pneumatic air-powered pump was installed that can be operated from the sidewalk that is approximately 130 ft above the base of the chamber. This protected against further clogged pumps, as well as power

Figure 10.6 View of the South anchorage sump pump enclosure. Photo credit: Port Authority of NY/NJ.

Figure 10.7 The 1-ton trolley hoist on a steel beam system erected in the anchorage. Photo credit: Port Authority of NY/NJ.

blackouts to the primary energy sources. In addition, a visual alarm system was installed to alert authorities when the anchorage was beginning to flood.

Providing support for the two pump systems and floats are structural steel channels spanning the pit and bolted to the anchorage floor on both sides. Currently, there are openings between the steel channels where debris can enter the pit. Pipe railing was installed on an 8 in. raised concrete curb that also serves as an additional method to secure the crossing steel channels. The discharge piping and pump bases are installed on raised concrete pedestals with gravel placed around the pedestals to contain sediment after the water is discharged into the pit.

Both sump pits have a steel trolley beam that is used during maintenance for removal and reinstallation of sump pumps. The trolley beam spans transversely across the sump room, as shown in Figure 10.7. When removing a pump, the pump is hoisted out of the pit vertically, and then it is rolled horizontally through a removable section of the railing to the center of the chamber for maintenance.

10.5 NEW JERSEY ANCHORAGE PUMPING APPARATUS

The south chamber of the New Jersey anchorage has the highest rate of water infiltration flow on a regular basis. The rate of flow varies on a fairly regular basis, but is well within the existing capacities of the pumps in use. Figure 10.8 shows water flowing into the pit. Table 10.1 shows the sump pump characteristics currently in place.

The sump pits in each chamber consist of duplex submersible pumping systems with float switch controls. This pumping system is designed to have 100% redundancy by alternating between the two pumps in each pit after each operating cycle. Each pump has a capacity of 150 gpm and 85 ft of head. The pumps are controlled by float switches arranged at predetermined levels within the pit to start the first pump, and then turn it off once the specified off level is reached. The pump activation will remove the collected ground-water inside of the pit and discharge it to the New Jersey storm water system. An additional benefit of alternating the pumps after each cycle of operation is to minimize wear and tear and continue to test the operational readiness of each pump. In cases when the incoming flow is higher than normal, one of the floats will activate a high-level alarm signal, which will be sent to the communication desk at the GWB administration building. If the incoming flow is such that the water level goes beyond the high-level alarm set point,

Figure 10.8 Water flowing into the New Jersey south anchorage sump pit. Photo credit: Port Authority of NY/NJ.

Table 10.1 Sump Pumps Located in Each Chamber of the New Jersey Anchorage

Apparatus	Capacity	Remarks
North sump pit	150 gpm	Duplex
South sump pit	150 gpm	Duplex
South and north chamber	65 gpm	Pneumatic diaphragm pump

Source: Quality Assurance Division Stage I Report, Internal document prepared by Port Authority staff. (QAD) Stage I Report.

another float will activate the second pump, and both pumps within the pit will run simultaneously until the water level within the pit reaches the off level. The controller for the pumps is located at the intermediate level of the chamber (approximately 30 ft above the pit) to protect the controller from flooding. The controller also monitors power on and pump off status, and it has a hand-off-automatic (HOA) selector switch. The status of each are sent to the communication desk for remote supervision.

Existing power to the New Jersey anchorage sump pump rooms is fed from an electrical room to the anchorage with an automatic transfer switch transferring between agencies as required.

In case of a blackout for the duplex pumping systems, the facility utilizes a backup pumping system consisting of a submersible compressed air–driven pump located at the bottom of the chamber in the vicinity of the sump pit, as shown in Figure 10.9. The compressed air hose extends from the pump casing up to the access entry point at the top upper-roadway level. The flexible hose is connected to a portable diesel-powered air compressor when the pneumatic pump operation is required, establishing an electricity-free backup system. This system can be activated from the sidewalk level should a flooding event occur.

The pumping systems are inspected and tested monthly for removal of foreign objects in the pit to prevent clogging, proper positioning of the level-sensing floats, control of pump noise and vibration level control during operation, safety conditions, testing of alarm indications at the communication desk and local panel, troubleshooting of electrical components, and power and control wiring condition.

Figure 10.9 The backup pneumatic pumping system. Photo credit: Port Authority of NY/NJ.

10.6 THE NEW YORK ANCHORAGE

Situated aboveground there is no infiltration of groundwater in the New York anchorage; however, corrosive elements have infiltrated the anchorage in the past through leaky roadway joints and through exposure to the elements. Previous steps taken in the 1970s diverted water from the leaky joints through a series of drainage troughs and piping. Through enclosing the anchorage, the cables have been protected from the elements.

10.7 DEHUMIDIFICATION

A condition that the PANYNJ continues to address is the high levels of relative humidity (RH) within the chambers. The levels of relative humidity are increased by the bridge proximity to a marine air environment and percolation of groundwater. These factors accelerate the rate of corrosion of the cables, which can compromise the structural capacity of the bridge. To ensure low acceptable levels of humidity and to prolong the life span of the bridge, the moisture in the air and ponding on surfaces inside the anchorage chambers are removed by a dehumidification system, as shown in Figure 10.10. The existing dehumidification system maintains each chamber below 40% RH year-round. In order to achieve this stable rate, each chamber contains two dehumidifiers connected in parallel and is equipped with reactivation air heaters for the reactivated (returned) air, internal bypass, dampers, and process and reactivation fans.

The dehumidifiers for each location connect to a duct network that spans from right below the upper-level roadway all the way down to the lower level of each chamber. This network discharges air through several process supply air registers. Air is drawn by the dehumidifiers from the top of the anchorage enclosure and reintroduced, post-dehumidification process, into the enclosure at the bottom of the anchorage. The dehumidification system is activated by the use of a dew point, relative humidity, and temperature sensors strategically located within the chamber cable area. The sensors transmit their signals to the direct digital controller, which activates the dehumidifiers. The dehumidifier will turn on when the humidity exceeds 40% RH and will continue to run until the humidity drops below that level. In addition, if the cable surface temperature is less than 10°F above the air dew point temperature, the dehumidifier will run until the air dew point temperature is 20°F below the cable surface. The power on, system run (manual or auto), and system fault status are locally monitored by indicator lights at the controller enclosure and sent to the watch engineer's office for remote supervision. The controller alternately runs the two dehumidifiers at each chamber on a weekly basis to minimize wear and tear.

Figure 10.10 The dehumidification system.

A key element of the dehumidification system is the use of a heavy-duty, high-quality polyester fabric membrane enclosing the main cables inside each chamber. The enclosure is used to create a closed and controlled environment around the critical parts of the anchorage to limit moisture infiltration and prevent dehumidified air from escaping.

The dehumidification systems are inspected and tested monthly to ensure that desiccant wheels, filters, seals, process air outlets, blowers; alarm indicators at the watch engineer's office and at the local panel; electrical components; and power and control wiring condition are in good condition and operating properly. In addition, the membrane dehumidification enclosure is inspected and maintained to ensure proper functionality.

To provide an extra level of reliability, both the sump pumping and dehumidification systems' electrical power is supplied with bistate power from both PSE&G (New Jersey) and Con Edison (New York City), with PSE&G being the primary source and Con Edison the emergency power source.

10.8 FUTURE MAINTENANCE

Projects in development include an additional measure for the New Jersey anchorage of a passive drainage system to allow removal of overflow water without the need for any pumping system. This will prevent any water buildup in the chamber.

The PVC piping currently in place has occasionally been damaged by contractors working within the anchorages. Those damages have required repair. Plans are in place to provide galvanized steel grating over many areas of the PVC piping system on the floor of the chamber. A future project will entail replacing the PVC piping with galvanized steel piping, which will minimize accidental damage.

To prevent clogging of sump pumps, a planned upgrade will utilize N-series submersible pumps, which allow small-diameter debris to be pumped through the system. These self-cleaning features reduce the risk of clogging and purport to be maintenance-free. This installation will include a complete new submersible pumping system at each chamber, consisting of duplex pumps, controls, discharge piping, valves, and associated structural and electrical modifications that enhance the dewatering capability and reliability.

Planned structural improvements include a railing access point to get into the pit. Steel gratings will be installed to provide a working surface over the pit to allow ease of access for pump maintenance and repairs are needed. Access hatches will be installed. The gratings will be designed to support the pump and live loads. Steel framing will be utilized to provide attachment locations for the sump pumps and floats, as well as to support the gratings and access hatches.

The following benefits will be realized with this future project:

- New equipment and piping for increased reliability
- Clog-free type N pumps that provide less susceptibility to clogging
- Liquid-level probe along with float switches for control and monitoring reliability
- Steel grating access panels that improve access and safety as well as reduce the chances of debris entering the pit
- Improved remote monitoring and alarm of pumping system operations
- Reduced maintenance costs

The estimated cost of these enhancements in 2014 dollars is approximately $900,000, with an estimated construction duration of 18 months.

10.9 CONCLUSION

Preserving the integrity of the critical structural elements George Washington Bridge anchorages is a challenge that the Port Authority of New York/New Jersey is managing through inspection maintenance and projects that improve the efficiencies of existing systems. Measures are in place to mitigate groundwater infiltration and to dehumidify the anchorages. Projects in development will improve the reliability of the existing dewatering and dehumidification systems and make them easier to maintain, improve monitoring of the dewatering and dehumidification systems and detection and notifications and malfunctions, and construct a backup, passive dewatering system that will not require pumps.

REFERENCES

1. George Washington Bridge across the Hudson River at New York. 1933. *Trans. ASCE*, 97.

Chapter 11

Rehabilitation of Suspension Bridges Supplemental Cables

David List and Charles Cocksedge

CONTENTS

11.1   INTRODUCTION

The Tamar Bridge forms a vital transport link carrying the A38 trunk road over the River Tamar between the county of Cornwall and the city of Plymouth in the southwest of England, and when opened in 1961, its main span was the longest suspension span in the country (Figure 11.1). The bridge is owned, operated, and maintained by its original sponsors, two local authorities, Cornwall Council and Plymouth City Council, and has relied solely on toll income to cover all capital and recurrent costs. The bridge is operated in conjunction with another estuarial crossing—the Torpoint Ferries—as a single business unit, by a joint committee formed by the two authorities.

In order to appreciate the context of supplemental cables on the structure, it is considered useful to first review its evolution.

Figure 11.1 Tamar Bridge.

11.2 THE ORIGINAL STRUCTURE

The original bridge was designed by Mott Hay and Anderson as a conventional suspension bridge with symmetrical geometry, having a main span of 335 m and side spans of 114 m, and with anchorage and approach spans, the overall length is 642 m (Anderson, 1965). Unusual for a suspension bridge of this era, the towers were constructed from reinforced concrete, and have a height of 73 m, with the deck suspended at half this height (Figure 11.2). The towers sit on caisson foundations founded on rock. Main suspension cables are 376 mm in diameter, and each consists of 31 locked-coil 60 mm diameter wire ropes, as shown in the cross section in Figure 11.3, and carry vertical locked-coil suspenders at 9.1 m centers.

The main cables are splayed at anchorages and anchored some 17 m into rock. The stiffening truss is 5.5 m deep and composed of welded hollow boxes, and the original deck, spanning between cross trusses, was of a composite construction with a 150 mm deep reinforced concrete slab on five universal steel beams. The deck was surfaced with hand-laid mastic asphalt 40 mm thick. The original bridge cross section is shown in Figure 11.4.

The main deck carried three traffic lanes, which were (and still are) operated in a tidal fashion—two lanes west and one lane east, or vice versa, to

Figure 11.2 Tamar Bridge during construction.

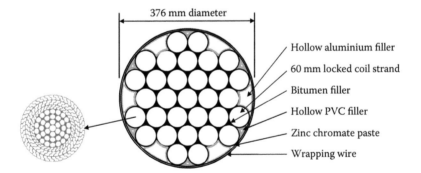

Figure 11.3 Main cable and locked-coil cross sections.

match the dominant traffic direction. A typical weekday traffic profile is shown in Figure 11.5.

Traffic flows have gradually increased during the life of the bridge, and by the late 1990s, the average annual two-way flow had reached 14 million vehicles. Since the mid-1980s, tolls have been collected from vehicles traveling eastbound only, and traffic data for this direction is shown in Figure 11.6.

Figure 11.4 Original bridge cross section.

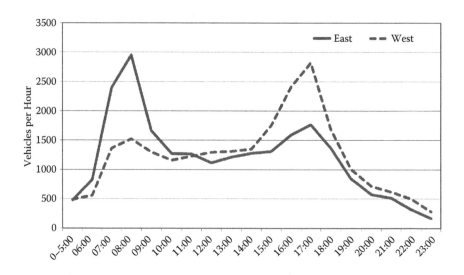

Figure 11.5 Typical weekday traffic flow.

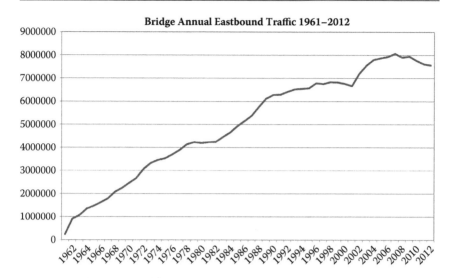

Figure 11.6 Annual tolled traffic since construction.

11.3 INCREASED LIVE LOAD REQUIREMENTS

A European directive (85/3/EEC) was published in 1985 requiring harmonization of vehicle weights and consequent increases in permissible weight of heavy goods vehicles traveling within member states. This increased the maximum weight of vehicles from 40 to 44 tonnes, and the maximum allowable axle weight from 10 to 11 tonnes, and the deadline for the UK's exemption from this legislation was 1999. This potential increase in axle weights would clearly have an immediate impact on live loads on the existing structure, and would influence fatigue characteristics in the longer term.

When the bridge was designed in the 1950s, the maximum permitted vehicle weight was 22 tons. At that time in the UK, bridges were designed for two types of loading: normal and abnormal. Normal loading is known as type HA and is in the form of an equivalent uniform lane loading plus a knife-edge load. HA loading intensity varied in accordance with the loaded length required to produce the worst effect. Abnormal loading comprises a single theoretical vehicle with two pairs of twin axles with four wheels per axle. The spacing of the axle pairs can be varied. This is known as type HB and is defined in units with one unit equal to 1 ton on each axle. The total weight of the vehicle is set between 120 and 180 tons (30 and 45 units), depending on the requirements for the bridge. Tamar Bridge used the heaviest weight of 45 units. In accordance with the design code applicable for the original design, BS 153 (BSI, 1954), either type HA or HB would be used separately.

This design loading standard remained in force until 1978, when a new standard was published, BS 5400 (BSI, 1978). This introduced limit state design principles and increased loading to reflect the heavier vehicles that were now using the UK's roads. Studies commissioned by the UK Department of Transport later that decade made use of weigh-in-motion data collected from various sites around the country, including the 988 m span Severn Bridge, and applied probabilistic techniques to create the required reliability in the loading (TRRL, 1986). This study concluded that the type HA loading contained in BS 5400 was still set too low for modern traffic, particularly for long-span suspension bridges, and a revised loading standard was issued, known as BD37/88 (UK DOT, 1988). This standard has effectively remained in force until the recent adoption of Eurocodes in 2011.

Under BD37/88, type HB loading had remained unchanged, except that HB vehicles should be considered to coexist with HA traffic, but with a 25 m gap at both ends with no loading. A lower load factor was applied when type HB loading was present, so that when considering elements designed by long loaded lengths (cables, towers, etc.), type HA loading alone is always the critical loading.

Figure 11.7 compares the load in a single lane using the 1950s design code (BS 153) and in the 1990s (BD37/88). It can be seen that between 250 and 500 m loaded length, the basic load intensity has approximately doubled.

As a result of the European directive, the various bridge owners in the United Kingdom undertook a nationwide program of bridge assessments, and in 1994, the Tamar Bridge underwent a comprehensive structural assessment, taking account of the increased axle loads and their potential effect on the structure over its forecast life. This identified that the primary structural components of the suspension system, anchorages, and towers could cope with the forecast loadings, although with limited reserve capacity. However, the 150 mm thick concrete deck was found to be significantly substandard, and inspections found many areas to be in a poor condition. Of particular concern were the cantilever deck slabs carrying the narrow footways, which could not support the weight of an errant truck, and the parapets were also substandard. It was clear that the concrete deck would

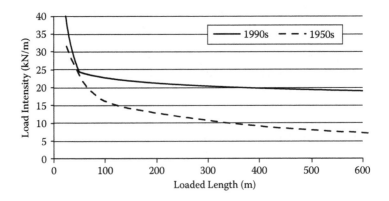

Figure 11.7 Comparison of lane loads from 1950s and 1990s.

have to be replaced. In addition, it was found that safety factors for some parts of the steel stiffening trusses would be unacceptably compromised by the increased loading.

A detailed inspection of the bridge had confirmed that there were no signs of distress, and no immediate weight restrictions were considered necessary, as the bridge did not carry a high percentage of heavy vehicles, and transit of abnormal loads was well controlled. However, it was clear that to comply with the European directive, action would be required in the medium term, as introduction of a weight restriction was not an option—while heavy vehicles (>3.5 tonnes) constitute less than 7% of bridge traffic, the crossing is a key component of the regional trunk road network.

11.4 FEASIBILITY STUDIES

The findings from the assessment meant that the upgrade work on the structure would be very significant and intrusive, and would become a milestone in the life of the structure. Therefore, rather than simply carrying out an immediate deck replacement and truss-strengthening exercise, it was decided to commission consultant Acer to undertake a broader feasibility study to recognize other ongoing concerns and issues. Consequently, in addition to reviewing the best options for strengthening, the study brief included the examination of other structural, operational, environmental, and aesthetic issues, including the following:

- Sag of the main span as a result of creep in the main cables
- Safety of users on the narrow footways
- Junction problems at both ends of the bridge
- Service disruption during the works
- Carriageway drainage
- Public and sustainable transport priority

11.4.1 Main Cable Profile

The bridge was designed to be built with a longitudinal profile hog of 3 ft (0.9 m) in the main span. However, on completion in 1961, this was in reality only about 2 ft (0.6 m), and by the mid-1990s, it had reduced to about 1 ft (0.3 m). The main reason for the original hog not being initially achieved was only discovered at the end of the strengthening and widening project, when it was found that the thickness of the reinforced concrete deck was somewhat greater than specified, leading to increased deck weight and consequential sag. The reason for the long-term sag was primarily due to the creep of the main cables and, secondly, due to the creep of the suspender cables (hangers).

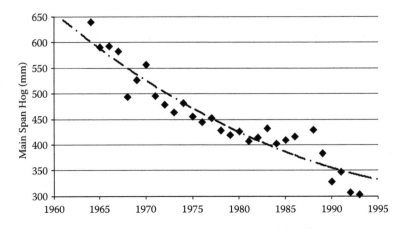

Figure 11.8 Reduction in hog of main span deck, 1964–1993.

The cables are all of the locked-coil type, which have layers of wires twisted around a central core wire. All of the wires are galvanized; that is, they are coated in a thin layer of zinc, which is relatively soft. Although the cables are subject to prestressing during manufacture to pull all the wires in the various layers tight, in service the continual tension causes the zinc under the crossing wires to flow slightly, and a slight straightening of the helical wires occurs. While the local effect is very small, over the full length of a 670 m long cable there was an increase in length of about 200 mm. This resulted in a lowering of the mid–main span of about 300 mm over the first 35 years of the structure's life. Comparison with other bridges in Germany and Canada with helical cables showed similar behavior.

One consequence of the creep that was noticeable was the associated effect on the wrapping wire and the inflexible paint system applied over it. The stretch in the cable had caused many circumferential cracks to open up in the paint.

Since the mid-1960s, an annual survey of the deck level had been carried out overnight, and this confirmed that the rate at which the cables were creeping was gradually decreasing, and the results are shown in Figure 11.8—the scatter of data points is considered to be largely a result of the difficulty in correcting for temperature effects, as the mid–main span level varies by about 10 mm per degree centigrade.

11.4.2 Options Appraisal

The owner required the feasibility study to investigate solutions that would give a high degree of certainty on cost, due to limited reserves and public sector borrowing restrictions—essentially, the project had to be funded from a combination of reserves and toll income received during the project period. This magnified the importance of minimizing traffic

impact. The bridge's three main deck lanes were at or approaching satura-
tion in peak periods, and any option that offered the availability of three
traffic lanes in peak periods during the construction period was going to
be very attractive.

A large number of options were generated and sifted. It was readily
apparent that if much additional dead load was added, then elements of the
suspension system could become overloaded. Therefore, the possibility of
replacing the concrete deck like for like was quickly ruled out because of
the need to increase the thickness to satisfy modern vehicle loadings, and
as a result, a steel orthotropic deck was proposed.

A series of options were generated, all of which increased the width of the
bridge deck. The more modest ones attached segregated footways outboard
of the truss top chords, whereas others attached a full lane. One of the
reasons for considering these was to use the additional deck width to provide the strengthening that was required for the truss top chords. If the
additional decks could be attached early on in the construction sequence,
then traffic could be diverted onto them to enable the contractor to readily
replace the degraded concrete deck while providing three traffic lanes to the
public. It was recognized that the significant additional weight of the cantilever lanes would further overload the structure, and this led to the more
detailed examination of options using supplemental cables—the idea of
adding supplemental cables had been conceived even before the feasibility
study had commenced, although initial thoughts had been largely notional.
The feasibility study examined potential layouts and rated their efficiency
in terms of the reduction in the amount of truss strengthening required. It
should be noted that strengthening the stiffening truss was not a straightforward matter, because addition of steel to provide the required increase
in strength also gives rise to an increase in stiffness—hence attracting
more load and a requirement for more strengthening material to be added.
Therefore, any means of reducing the requirement for truss strengthening
by adding, for example, supplemental cables was seen to be the most cost-
effective and efficient way forward.

11.4.3 Supplemental Cables

While it may appear that the supplemental cables are arranged and function like those on a cable-stayed bridge, this would be a misguided over-
simplification. In order to recognize this, it is necessary to understand the
articulation of the bridge (Figure 11.9).

The bridge is tied back to the anchorages at both ends via the approach
spans, which are fixed to the tops of the side towers. Thus, the side towers flex
with changing temperature. The cable saddles at the side towers are mounted
on rollers, and therefore, saddles can move independently. Each of the three
suspended spans is supported on rocker bearings at their ends, with those at the
main towers doubly articulated. The side spans are tied back to the approach

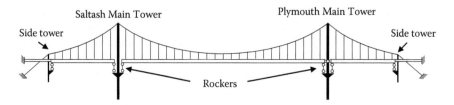

Figure 11.9 Bridge articulation.

spans, and the main span was tied to the Plymouth side span via a universal joint at the Plymouth main tower. Therefore, all temperature expansion and contraction takes place at the joint at the Saltash main tower. Other joints were present to permit rotation of the three suspended spans and to break the continuity of the concrete deck such that it did not participate in global action.

The layout of the supplemental cables was strongly influenced by practical consideration of suitable anchorage points, and this was particularly important at the tower tops. Unusually for a bridge of its era, it has concrete towers with a conventional cast-steel saddle bolted to the tower top. During bridge construction, the position of the saddle was adjusted by jacking it along a slide path comprising steel beams incorporated into the tower top. This ensured that the towers were vertical on completion, with the horizontal component of the cable forces in balance. It was fortunate that the cable saddles only occupied the central portion of the tower top, leaving convenient areas on either side as potential locations for anchorages for the supplemental cable system (Figure 11.10).

Preliminary cable arrangements had four cables fanning out from each tower top, with a pair each side of the main cable saddle. In the main span,

Figure 11.10 Cable saddle on main towers.

Saltash Main Tower   Plymouth Main Tower

Figure 11.11 Initial supplemental cable layout.

two cables were to be attached to the same truss node point, but in the side spans, one picked up the midpoint of the side span truss and the other was taken down to the base of the side tower to provide a tieback to the tower top. To provide a continuous cable system, horizontal cables would run from the middle of the side spans to the anchorages, and a cable at the bottom chord level was provided between quarter points in the main span (Figure 11.11). Essentially, the new cable system effectively formed a secondary catenary, albeit with very few suspenders.

It was recognized at this early stage that to optimize appearance, the new inclined cables must lie below the main cable catenary. This kept the cables at a reasonable angle to the deck to promote efficient load transfer; however, it meant that the cables could not extend far into the center of the span.

As the design developed, it soon became apparent that the permanent addition of cantilever lanes offered a cost-effective improvement in terms of both capacity and safety, as well as satisfying the temporary diversion requirement, although such permanent widening would require primary legislation—a new act of Parliament—as the additional landfall and air space required were unsurprisingly not within the scope of the original 1957 Tamar Bridge Act. A comparison between the cross sections of the original and the widened structure is shown in Figure 11.12.

11.4.4 Associated Investigations and Inspections

While there was well-founded confidence in the condition of the majority of the bridge structure, it was not known if the main cable was still in a good condition, given emerging results from internal cable inspections being carried out in the United States. Therefore, the owner determined to carry out a limited investigation and decided to remove the wrapping wire and paste from one of the mid–main span panels to examine the surface condition of the outer strands. The findings were very encouraging, as after cleaning, all strands were found to be in an excellent condition, with all of the zinc galvanizing intact and no sign of brown rust (Figure 11.13). This gave the owner more confidence that investing in the existing bridge was the correct decision, rather than contemplating more expensive alternatives, such as a replacement crossing.

Figure 11.12 Bridge cross section before and after strengthening and widening.

Figure 11.13 Main cable after unwrapping 1998.

Another investigation on the existing cable system consisted of carrying out a fatigue test on one of the suspenders to determine the remaining life. With the bridge approaching its 40th birthday, it might have been expected that suspender replacement would be required in the near future. An inspection of the suspenders had identified that one of the shortest at mid–main span was showing significant corrosion as the cable entered the lower socket, and the owner determined to remove it and carry out a fatigue test to failure. Although the suspender spacing is only 30 ft (9 m), the stiffening truss could not tolerate a missing suspender, so a temporary one was designed and installed. It was planned to provide a permanent replacement during the strengthening and widening contract.

In a suspension bridge, the suspenders are subject not only to a variation in axial load, but also to bending effects arising from differential cable and deck movements from unsymmetrical live load and wind load. These produce greatest stresses in the shortest suspenders at mid–main span, and so the fatigue tests comprised both axial tests and axial plus bending tests. Again, the findings were encouraging and showed that the existing suspenders had at least 120 years life remaining. The proposed supplemental cable system would improve the life of the suspenders by reducing the traffic loads carried by them and, more importantly, by reducing the longitudinal bending effects.

In order to gather information on vehicle weights crossing the bridge, a weigh-in-motion system was installed at the west end of the structure in all three traffic lanes. The tolling process collected basic statistical traffic information, but only in the tolled (eastbound) direction. The intention of the weigh-in-motion system was to collect data useful in the design, such as fatigue spectra, and to estimate the likely level of stress locked in during the work while traffic continued to cross the bridge. In order to achieve economy, the weigh-in-motion data was used to develop traffic loadings, which could be safely adopted during the implementation of the strengthening and widening works.

11.5 CONTRACT PROCUREMENT

Having appointed designers in 1996, the first phase of the design developed the feasibility study and gave a cost estimate of between £25 million and £30 million, stretching the client's budget to the limit, and highlighting the need for maximization of outturn cost certainty.

This need for cost certainty, coupled with a desire to maximize the scope for innovation from the designer, contractor, and client, led to the adoption of a partnering approach for the project. This involved tendering for construction with only part of the design fully detailed, and offering contractors the flexibility to take on all or any of a range of risks in their tenders (e.g., weather risk). Bids were invited on a quality and price basis, with tenderers essentially pricing only for the orthotropic deck. Cleveland Bridge of Darlington, United Kingdom, was appointed as the preferred contractor in March 1998, and Cleveland Bridge staff joined the project team. Early contractor involvement optimized opportunities for innovation and value engineering. In parallel, the necessary primary legislation had been pursued, and the Tamar Bridge Act 1998 gained royal assent in July 1998.

Active team building took place to form a strong partnership among the client, consultant, and designer, employing the services of a professional facilitator to lead a series of partnering workshops. The strength of this partnership would be well tested in the course of the project.

The contract form provided for the contractor's submission of a target price, which required the contractor's appraisal of the existing partial design and developing a comprehensive scope of works, working closely with the consultant and client. A target contract price of £23.5 million was subsequently agreed on and accepted, with contract signing in March 1999, following completion of legal arrangements, and work started immediately thereafter. This target price gave an overall project budget estimate of £31.6 million.

The contract was structured to financially bond the client, contractor, and consultant, by sharing of any savings or additional costs, but with an upper bound above which the contractor carried all additional cost. This financial structure would strongly influence all parties in due course. Further details of this project are given in the paper by List (2004).

11.6 DESIGN DEVELOPMENT

The contractor had been chosen on the basis of a fixed-price element—the replacement steel orthotropic deck—which was fully designed and detailed, as this element represented a significant proportion of the contract price. However, other parts of the design had not been fully developed, and it was intended to make use of the contractor's particular expertise and knowledge to help create the best solutions. Nevertheless, the designer retained full responsibility for the design, even if it was not the originator of the idea.

One important area where the contractor brought its expertise was the supplemental cable system. Cleveland Bridge had constructed a large number of suspension bridges, including the Tamar Bridge, and had a strong established relationship with the cable supplier Bridon, and proposed it as provider of the new cables. Bridon had supplied the original locked-coil strands for the main cables and suspenders 40 years earlier.

The original designer of the bridge had chosen to use locked-coil strands for both the main cable and the suspenders. The main reason for this is understood to be experience following the dismantling of an old transporter bridge where the spiral strand cables were found to have significantly deteriorated, and it was considered that the smooth exterior surface provided in a locked-coil strand would provide superior long-term corrosion protection. The soundness of this decision was certainly validated by the inspections and testing after 40 years of service. For consistency and based on excellent prior performance, it was decided that the supplemental cables would also be of locked-coil construction.

The main aim of the design development was to bring all of the design up to a stage at which the contractor could produce the target price for acceptance by the client. Final detailing by the designer would continue in accordance with a schedule agreed to by all parties.

Further work was carried out by the designer to optimize the cable layout, and this produced some modifications to the original plans. In the main span, the pair of supplemental cables from each tower top had been originally designed to pick up the stiffening truss at the same node, but it was determined that it would be more effective and efficient to pick up two different nodes approximately 27 m apart. In the side spans, the modified design retained one cable from each tower, picking up the midpoint truss node, but instead of a continuity cable extending back to the anchorage, an easier solution was developed that made use of the cantilever decks to carry the tension back to the anchorage. The final supplemental cable arrangement is shown in Figure 11.14. The cantilever decks were tied to the anchorage using 50 mm diameter prestressing bars (Figure 11.15).

The other cable in the side spans runs directly from the tower top down to the bottom of the side tower. This concept was unchanged from the original design, although careful geometrical checks were made to ensure it did not foul any of the struts supporting the new cantilever decks. To carry the horizontal component of the cable force to the anchorages, posttensioned concrete beams were designed to link the bases of the side towers to the anchorages (Figure 11.16).

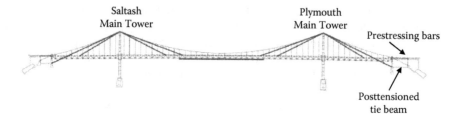

Figure 11.14 Final supplemental cable layout.

Figure 11.15 Cantilever deck tieback.

The connection details of the supplemental cables to the truss and towers were developed considerably during this period. The contractor expressed a strong preference to jack the cables from the bottom ends, as it would simplify handling of the heavy jacking equipment, so at the tower tops, a simple arrangement was possible with open (spelter)-type sockets fixed to a common bracket fixed down to the tower top. The critical design case for the tower-top bracket catered for one of the cables being accidentally

Figure 11.16 Posttensioned tie beams.

severed, which would produce a large out–of-balance load, potentially supplemented by dynamic effects. However, there was a separate legacy issue regarding a series of cracks in the concrete below the saddles that were visible on the inside surfaces, and to deal with these, it was proposed to drill and insert prestressing bars horizontally through the concrete below the saddles in both longitudinal and transverse directions. To assist in resisting the cable loss design case, it was decided to make use of the remedial prestressing bars and connect their bearing plates to the supplemental cable brackets. As a final measure, the cable brackets were butted against and welded to the main cable saddle base plate. The tower-top cable brackets and posttensioning are shown in Figure 11.17.

At the truss end of the supplemental cables, where jacking was to take place, a detail was developed where a 30 mm thick fin plate is attached to the corner of the top chord box member. This suited the construction of the box, which had full-height web plates, so the fin plate could be butt-welded directly to the top of the web. The new cables had block sockets attached at the lower ends, with twin anchor bars attaching to a crosshead that was pinned to the fin plate. While this was not the most elegant solution, it provided a convenient method for jacking and offered the facility of significant length adjustment (Figure 11.18). This length adjustment was considered necessary as, although dimensional surveys of the bridge had been carried out, the bridge was continually moving under traffic and thermal effects.

Figure 11.17 Tower-top cable bracket and prestressing bars.

Figure 11.18 Supplemental cable—tower and deck connections.

Figure 11.19 Supplemental horizontal cable under truss.

To provide longitudinal continuity along the middle of the main span, cables were placed below the bottom chord within an inverted-U-shaped structural member that was provided to strengthen the chord (Figure 11.19).

11.7 CONSTRUCTION

The overall construction sequence was developed around the needs of the structure and the need to keep traffic running across the bridge in both directions. This required the cantilever lanes to be installed so that the traffic could be transferred onto them, to permit the substandard concrete deck to be replaced. In outline, the sequence was as follows:

- Preparatory work, including carrying out diversions of gas and water pipes at the ends of the bridge, and building of scaffolding beneath the entire length of the stiffening truss
- Truss strengthening
- Supplemental cable installation and stressing
- Cantilever deck installation from all four "corners," working symmetrically toward center
- Cantilever deck surfacing and transfer traffic off central deck to cantilevers
- Replacement of westbound half of central deck
- Replacement of eastbound half of central deck
- Central deck surfacing
- Removal of temporary works and scaffolding
- Final adjustment of supplemental cable tensions

The traffic management arrangements were designed to provide three running traffic lanes through all peak periods—typically weekdays 0700 to 0900 and 1500 to 1830. Outside these periods, the contractor was generally permitted to reduce the public facility to two traffic lanes. A free shuttle bus for pedestrians, mobility scooters, and cyclists was operated 24/7 throughout the works. Twelve overnight full-bridge closures were permitted to allow some specific activities to be undertaken, such as installation of tower-top elements and temporary works. During these overnight full closures, a free passenger ferry was provided for pedestrians, and special arrangements were put in place to facilitate controlled crossings by "blue light" emergency vehicles. The traffic management for the various construction phases is shown in Figure 11.20.

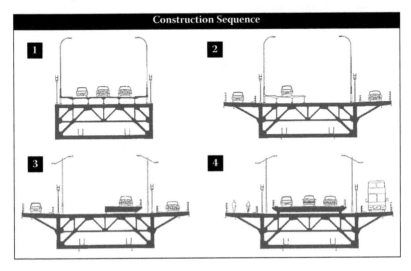

Figure 11.20 Traffic management.

The supplemental cable sizes were rationalized to two diameters—102 and 112 mm—to create economic lengths for manufacture. These had minimum breaking loads of 10,100 and 12,200 kN, respectively. The design of the cable construction and the sockets and fittings was carried out by the cable manufacturer, Bridon. Details of the strand construction are given in Tables 11.1 and 11.2. Illustrations of a typical locked-coil strand are given in Figure 11.21.

In order to prove the design, tests to destruction on specimen lengths of both diameters were carried out. These tests found the designs to be satisfactory, with each specimen failing just above the required minimum breaking loads at 10,435 and 13,000 kN for 102 and 110 mm diameter, respectively. The load–extension plot for the test of the 102 mm diameter cable is shown in Figure 11.22. The initial stages of the test are bedding in cycles and are used to determine the effective Young's modulus. This is followed by the test to failure. The downward spike near the failure point records when some of the outer Z-shaped wires failed first—it was noted that these have a somewhat lower tensile strength.

Table 11.1 Strand Design Data

	Unit		
Nominal cable diameter	mm	102	112
Guaranteed minimum breaking load	kN	10,100	12,200
Modulus of elasticity	kN/mm^2 ± 5%	155	155
Steel area	mm^2	7,074	8,701
Nominal turn factor	deg/kN/m	0.000	0.000

Table 11.2 Strand Construction Data

	102 mm				112 mm		
Type of Wire	Wire Size (mm)	No. of Wires	Tensile Strength Range (N/mm^2)	Type of Wire	Wire Size (mm)	No. of Wires	Tensile Strength Range (N/mm^2)
Full lock	6.00	56	1520–1770	Full lock	6.00	62	1470–1720
Full lock	6.00	49	1520–1770	Full lock	6.00	55	1520–1770
Full lock	6.00	43	1520–1770	Full lock	6.00	48	1520–1770
Full lock	6.00	36	1520–1770	Full lock	6.00	42	1520–1770
Round	5.00	29	1770–2020	Full lock	6.00	35	1520–1770
Round	5.00	23	1770–2020	Round	5.00	28	1770–2020
Round	5.00	17	1770–2020	Round	5.00	22	1770–2020
Round	4.80	12	1770–2020	Round	5.70	14	1570–1820
Round	2.85	6	1770–2020	Round	3.50	7	1770–2020
Round	3.70	6	1770–2020	Round	4.60	7	1770–2020
Round	3.45	1	1770–2020	Round	4.60	7	1770–2020
				Round	6.60	1	1570–1820

Figure 11.21 Typical locked-coil strand.

The cables were made at Bridon's factory in Doncaster, England. Initially, the wires were drawn and galvanized, having gone through lead patenting heat treatment, and then the cables were made in the adjacent ropery, with each layer being subsequently built on top of previous layers. The layers generally alternated in direction to create a balanced design, with the same area wound clockwise and anticlockwise, removing the tendency for the cable to twist under load. A proprietary blocking compound was used during strand assembly to fill all spaces between the wires.

The completed strands were then taken to the prestressing and socketing area. For each cable, a permanent socket was fitted to one end and a temporary socket at the other end. The cable was then run out along the prestressing track and tensioned cyclically up to half of its breaking load in order to bed the wires in and remove the nonlinear load–extension characteristics. The cable was then loaded to its "marking length," which is its intended dead load tension, and the position of the second permanent socket was marked. The cable was then removed from the track for the second socket to be fitted. The procedure for socketing the strands is as follows, illustrated in Figures 11.23 and 11.24.

- Wrap wire tightly around the part of the strand at the mouth of the socket (known as serving wire).
- Insert the strand into the socket through the mouth.
- The wires within the socket cone are individually brushed out to occupy the space.
- The socket and strand are lifted into a vertical position.
- The socket is preheated and molten zinc is poured into the cone.

The completed cables were wound onto large-diameter spools for transport and unreeling on site—the large diameter of the spool limited curvature and ensured that individual wires did not pop out of the strands. A typical spool is shown in Figure 11.25.

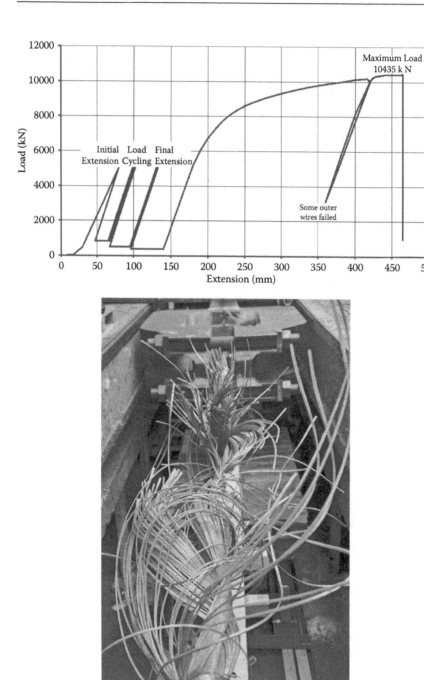

Figure 11.22 Load–extension plot for test and broken cable.

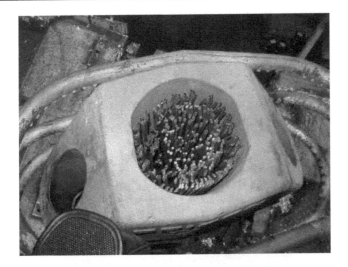

Figure 11.23 Wires brushed out.

Figure 11.24 Application of preheat.

The erection of the cables required extensive temporary works. The key elements were the platforms erected on the tower tops. Their main aim was to support the winching sheaves, but they also served other purposes, such as serving as drilling platforms for the posttensioning to contain the cracks in the concrete.

Winches were positioned on the ground on both sides of the river in convenient positions to keep winch line routes as simple as possible. Strand handling operations are illustrated in Figure 11.26.

Figure 11.25 New cables delivered on spools.

All of the inclined cables attached to the tower tops were tensioned in the same way. The upper sockets were attached first to the tower-top brackets via pins through the spelter sockets, and then the lower block sockets were fed over the jacking rods at the bottom. The permanent rods were provided with extension rods to accommodate the hollow ram jacks and provide sufficient movement to pull the cable from slack to its full tension. The jacks were seated against the top of a fabricated T, with the stalk bearing against the top of the block socket. This permitted the cable to be progressively jacked into position (Figure 11.27).

The midspan cable below the bottom chord required different techniques. This had to be pulled off its reel down onto the scaffolding platform under the deck, and then lifted up inside the bottom chord strengthening member. These cables were provided with face-bearing cylindrical sockets, one of which had an internal thread at the back to permit a draw bar to be attached for jacking. To provide space for the jacking equipment, a section of the strengthening member had to be temporarily removed, as well as the bottom chord in this area locally strengthened. The jacking arrangement is shown in Figure 11.28. As the cables were jacked, shims were placed against the front face of the socket to maintain the extension. To prevent the cables sagging, a series of small saddles were provided along the length of the strengthening member with a low-friction seating material.

Figure 11.26 New cables being hauled to tower top and handled on deck.

On completion of installation and jacking, a protective coating system was applied to the supplemental cables. Bridon's Metalcoat system had performed well on the suspender cables and was chosen for the supplemental cables. This product is a suspension of aluminum flakes incorporated into a hydrocarbon carrier, diluted with solvent for ease of application. While the surface becomes touch dry, the coating remains flexible, allowing for differential wire movements in service, eliminating surface cracking.

Figure 11.27 Cable jacking at live (truss) end.

11.8 CONSTRUCTION MONITORING

In view of the long-term sagging of the main span that had been noted since the bridge opened, a level monitoring system had been installed in 1994 in the main span. However, this had not proved to be very reliable, and an annual level survey using traditional techniques had continued.

As part of the strengthening and widening project, it was considered that a structural health monitoring system should be included. In line with the budgetary constraints of the project, it was agreed that the new system should only incorporate features that were absolutely necessary. After a review of proven available systems, it was decided to replace the electrolevel

Figure 11.28 Cable-jacking frame for underdeck cable (top) and with jacks (bottom).

system with a liquid-level sensing system across the main span. For the supplemental cables, strain gauging of the anchor bars was chosen as a simple and reliable method. Temperature sensors were fitted to the suspended structure, and a new environmental monitoring system was installed to provide the bridge control staff routine information on temperature and wind speed and direction. The new structural health monitoring system was installed at the beginning of the site work.

Figure 11.29 summarizes the changes in the level at mid–main span during the strengthening and widening construction works. The graph shows both predicted (analysis) and measured midspan deck level. The

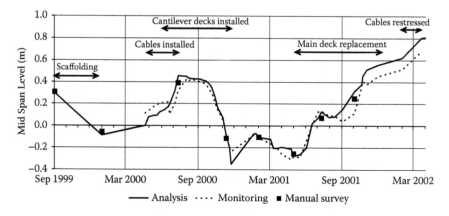

Figure 11.29 Variation of deck level at mid–main span during construction.

stage-by-stage analysis covered over 50 steps, including the addition and movement of temporary works and scaffolding. The principal activities are highlighted over the periods when they took place. The initial stages of adding scaffolding underneath the whole length of the bridge caused the main span to just sag below horizontal, but after the new cables were installed and stressed, the deck was lifted up to above 400 mm. The erection of the cantilever decks slowly caused the level to reduce again, and it ended up significantly sagging through the addition of 1000 tonnes for each of the new cantilever decks. This was inaccurately picked up by the local newspapers, which reported that the bridge was sinking, and required a careful explanation that the bridge was sagging as predicted and was still safe for traffic to cross. As the concrete deck was gradually replaced by the lighter steel orthotropic deck, the midspan level progressively raised.

Table 11.3 summarizes the contributions of various elements of the structure to the suspended weight, before, during, and after the works.

The change in shape of the deck longitudinal profile at various stages is illustrated in Figure 11.30.

Table 11.3 Summary of Weight of Suspended Structure

	March 1999 Start	March 2001 Heaviest	On Completion
Cables	900 t	1,025 t	1,025 t
Steel	2,400 t	4,425 t	5,200 t
Concrete	3,500 t	3,500 t	0t
Surfacing	1,100 t	1,700 t	1,700 t
Scaffolding	0 t	650 t	0 t
Total	7,900 t	11,300 t	7,925 t
Midspan hog	+300 mm	−200 mm	+800 mm

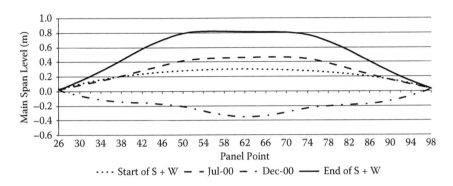

Figure 11.30 Change in main span longitudinal profile during construction.

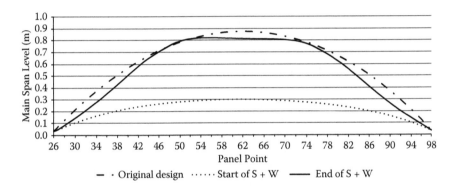

Figure 11.31 Main span longitudinal profile, as designed, at start and end of strengthening and widening project.

It is apparent that at the final stage, there is a distinct flat region across the middle third of the span. This is a consequence of the necessarily compromised locations of the attachment points of the new cable system not extending to near the middle of the span. However, the overall profile was reasonably close to the original design profile, and this was considered by the design team to be an excellent result, as shown in Figure 11.31.

In order to improve the appearance and drainage of the deck at the center of the span, the new main deck was installed on stools of varying heights to give a good-looking profile. The final profile is shown in Figure 11.32. It was not possible to do the same with the cantilever decks, as they were welded to the top chords, but the cantilever struts were given additional packing to lift the edge and fascia beam to form a smooth curve. This gave a slightly increased crossfall to the cantilever decks local to midspan.

Other geometrical changes became apparent as the project progressed; the most significant of these was a longitudinal movement of the deck toward both anchorages, resulting in an opening of the gap at the main expansion

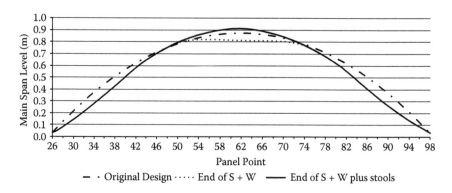

Figure 11.32 Final longitudinal profile.

joint at the Saltash main tower. Part of this movement was confirmed to be attributable to the supplemental cable system producing a longitudinal compressive stress in the stiffening truss, but this was enhanced by what was believed to be weld shrinkage caused by the large amount of welding carried out through the attachment of the cantilever decks and other miscellaneous strengthening steelwork. This became apparent when the expansion joints for the cantilever decks were installed and the gap was larger than expected. A simple solution was possible through the use of a longer bridging plate across the gap. As the new main deck around the joint had not been fabricated, it was possible to make the units slightly longer, so that the originally envisaged joint could be fitted. The theoretical gap at the Saltash main tower throughout the strengthening and widening project is plotted in Figure 11.33.

This longitudinal shrinkage produced another noticeable effect, in that the suspenders no longer remained vertical to varying degrees. This was particularly apparent with the shortest suspenders at mid–main span, where they were out of plumb by about 80 mm. There were initial concerns about the effects this would have on the suspenders ropes, but this was a static effect that has simply changed the static stress in each wire. The test carried out on one of the

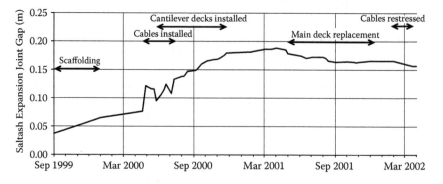

Figure 11.33 Change in expansion joint gap at Saltash main tower.

suspenders before the works had showed ample remaining fatigue life, particularly as the supplemental cable system reduced the variation in axial and bending stresses. A plot of the lean of the mid–main span suspenders over time is shown in Figure 11.34—unsurprisingly, virtually mirroring the history of the expansion joint gap. Figure 11.35 shows a leaning suspenders near midspan.

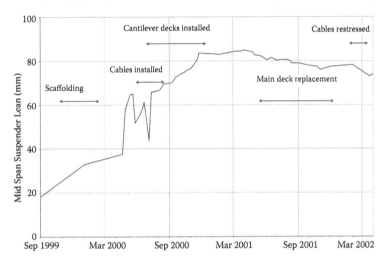

Figure 11.34 Change in verticality of suspender at mid–main span.

Figure 11.35 Nonvertical suspender near mid–main span.

11.9 IN-SERVICE BEHAVIOR

All of the supplemental cables had been fitted with strain monitoring instrumentation, which could be analyzed and presented as loads on the structural health monitoring interface. This data is presented as 24 h history in graphical format, together with current loadings. An example partial screen shot for a pair of cables (north and south P2) is shown in Figure 11.36.

This information has been used to monitor the behavior of the structure both under normal day-to-day operation and in special circumstances, such as the transit of exceptionally heavy vehicles, for example, >300 tonnes. It has also been analyzed in conjunction with dimensional analysis.

Soon after installation and tensioning, it became apparent that some of the supplemental cables were susceptible to rain and wind vibrations and oscillations. Observations suggested that the amplitude was generally only around 100 mm, and these occurred with a wind speed of between 10 and 15 m/s and lasted up to 90 min. In all cases, it was known to have been raining, and it was concluded that the cause was a phenomenon widely known in the cable-supported bridge fraternity, where a rivulet of rain running down the underside of the cable distorts the airflow, producing vortex shedding and consequent vibration. Compared with other bridges suffering from this effect, the magnitude at Tamar was quite modest, and so not of immediate concern.

Before committing significant funds to designing and installing proprietary dampers, a very simple solution was trialed using a dashpot-type damper. This used a water butt located on the deck as the cylinder, and a rod clamped to the cable connected to a circular paddle acting as a piston inside the water butt. Following installation, vibrations immediately ceased, confirming that sufficient damping was available. These dampers require no maintenance, with the damping fluid being automatically naturally replenished by rainfall (Figure 11.37). The pure economy and efficiency of this solution has attracted some very favorable comments from highly respected bridge engineers, and this has curtailed any aspirations for a more elegant arrangement.

As part of a research project, the University of Sheffield was permitted to install some vibration monitoring equipment on the bridge. This included the provision of accelerometers on some of the supplemental cables. As part of the commissioning tests, the effectiveness of the hydraulic dampers was assessed as shown in Figure 11.38 (Koo et al., 2013). The damping

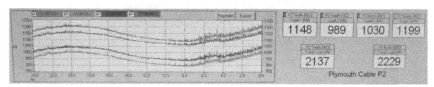

Figure 11.36 Monitoring system output.

Figure 11.37 Hydraulic cable damper.

provided is very effective in the vertical direction when compared to the horizontal direction, where there is no additional damping provided. It can be seen that the time taken for the vibration to fully decay reduces to about one-eighth in the damped direction.

11.10 INSPECTION, MAINTENANCE, AND IMPLICATIONS OF SUPPLEMENTAL CABLE SYSTEM

Inspection of the structure is based on the UK Highways Agency's *Design Manual for Roads and Bridges* BD63/07 (UK DOT, 2007), which has a 6-year cycle for principal inspections, but uses a risk-based approach to prioritize elements within the rolling 6-year cycle. The inspection regime comprises

- Ad hoc superficial inspections
- General inspections every 4 months
- Annual inspections focusing on specific elements of the structure, for example, surfacing, movement joints, bearings, and so forth
- Special inspections as required, for example, internal main cable inspections, suspender bolt inspections, permanent access gantries, and so forth

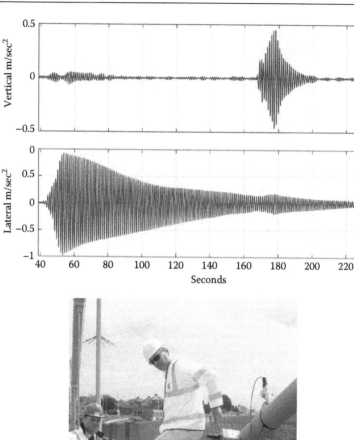

Figure 11.38 Vibration test of supplemental cables.

An internal inspection of the main cables was carried out in 2008 of two panels on the south cable. Easily accessible low locations were chosen at the middle of the main span and at the end of the Saltash side span—these were also judged to be potentially the worst locations where water was known to run down and collect. Previous inspections had just removed the wrapping wire and cleaned off the paste so that the surface of the outer strands could be viewed. However, for the 2008 inspection, it was decided to probe deeper, and so the strands were wedged apart to enable the outer surfaces of the inner strands to be inspected. No attempt was made to open up any of the locked-coil strands, as it was considered that this would only cause unnecessary damage. The gaps between the strands had been filled with a proprietary bituminous compound that was found to have performed very well, such that after cleaning off, the exterior of the strands were in an excellent condition, with only very occasional minor patches of brown rust visible (Cocksedge, 2009). It was concluded that no intervention, such as dehumidification, was required, or the installation of acoustic monitoring. Separately, dehumidification was installed in the four anchorage chambers in 2009, where more extensive corrosion of the individual strands had been identified. The chambers had been known to collect water from groundwater penetration through construction joints in the concrete walls, and the reduction in humidity has prevented the reappearance of any corrosion.

A medium-term maintenance program has been determined for the period 2014–2020 that includes the following:

- Improved permanent bridge access arrangements
- Phase 1 of the bridge recoating strategy (2–4 years duration)
- Bridge resurfacing—main deck main span and west side span and north cantilever
- Main cable hand strand replacement
- Main cable and supplemental cable maintenance and recoating
- Suspender recoating
- Replacement of main deck movement joints
- Replacement of the weigh-in-motion system
- Replacement of the structural monitoring system
- Investigation of rocker/pendel movement and wear
- Curb drainage unit replacement feasibility study and review
- Review of permanent access gantry runway beam supports
- Internal main cable inspections

The supplemental cables introduce additional inspection and maintenance tasks, but they are reasonably accessible and, for the most part, visible, and do not represent a particular burden in relation to the structure as a whole. The proprietary Metalcoat protective coating has performed very well, and no remedial work or recoating has been required in 14 years of service.

The addition of the supplemental cable system does offer some longer-term benefits beyond the relief offered to other elements of the structure under normal loading. The history of the main span profile during the strengthening and widening project demonstrated the facility to use jacking of the supplementary cables to influence the main cable catenary and the deck profile. This facility may be used in the future to offset future creep in the main cable and suspenders.

The facility to adjust the loading in the supplemental cable system may also contribute to decisions on other maintenance and improvement initiatives. For example, the east side span of the main deck was resurfaced with 55 mm thick Gussasphalt surfacing in 2011—some 25% thicker and heavier than the previously used 40 mm thick mastic asphalt—adding dead load to the structure. Based on performance to date and monitoring of the resulting reduced stresses in the orthotropic deck, this thicker surfacing may be an attractive option for the main span, where the additional load might contribute to a need to jack the supplemental cable system to share loading and offset sag.

11.11 CONCLUSIONS

The supplemental cable system was a key element in a design that enabled the owner to procure an efficient and effective solution to upgrade the Tamar Bridge in response to a European directive and its associated potential increase in the design weight of traffic, with the bridge having been assessed as understrength in a number of areas.

The supplemental cables were able to perform a number of functions, including permitting the proposed widening scheme to be implemented in the most efficient manner, reducing the amount of strengthening required, and returning the bridge profile to near the original design.

Although it was not an issue for the Tamar Bridge, the supplemental cables have improved the safety factor for the main cables. While not currently of particular importance, this may prove to be valuable in the future to deal with issues such as cable deterioration or further increases in loading conditions.

The cable layout was adapted to suit the opportunities available in terms of details at the tower tops, truss connections, and for tying back to the anchorages. This solution allowed the bridge, a vital link in the regional transport infrastructure, to remain open during the works, with minimal impact on the traveling public, and has increased the capacity and extended the service life of the structure.

The new cables were found to be susceptible to rain and wind vibrations, but oscillations have been eliminated through the provision of simple hydraulic dampers, showing that uncomplicated solutions can be achieved. Although they introduce some additional inspection and maintenance work, this has not been found to be demanding in relation to the structure as a whole.

In addition to its primary purpose as a feature of the strengthening and widening project, the cable system also offers a further structural monitoring opportunity and a new facility to undertake some limited adjustments to its load share. This offers the owner a tool to make limited adjustments to the geometry of the structure, responding to the future effects of time and loading changes.

REFERENCES

Anderson, J.K. 1965. Tamar Bridge. *Proceedings of the Institution of Civil Engineers*, 31(4), 337–360.

British Standards Institution (BSI). 1954. Girder bridges: Part 3—Loads and stresses: Section A—Loads. BS 153.

British Standards Institution (BSI). 1978. Steel concrete and composite bridges: Part 2—Specification for loads. BS 5400.

Cocksedge, C. 2009. No cause for concern. *Bridge Design and Engineering*, 55.

Koo, K.Y., Brownjohn, J.M.W., List, D.I., and Cole, R. 2013. Structural health monitoring of the Tamar suspension bridge. *Structural Control and Health Monitoring*, 20(4), 609–625.

List, D.I. 2004. Rejuvenating the Tamar Bridge: A review of the strengthening and widening project and its effects on operations. Presented at the Fourth International Cable Supported Bridge Operators' Conference, Copenhagen.

UK Department of Transport (UK DOT). 1988. Loads for highway bridges. BD37/88.

UK Department of Transport (UK DOT). 2007. Inspection of highway structures. BD63/07.

Transport and Road Research Laboratory (TRRL). 1986. Interim design standard: Long span bridge loading. Contractor Report 12.

Chapter 12

Suspension Bridge Security Risk Management

Sreenivas Alampalli and Mohammed M. Ettouney

CONTENTS

12.1 INTRODUCTION

Recent devastating security hazards brought to focus the need for bridges to implement measures to reduce their harmful effects. Implementing many such measures is very expensive, whether physical or operational measures, and could pose a major problem to bridge owners. Budget limitations in the current environment make this problem even more critical. Due to their symbolic and iconic nature, as well as the high volumes of traffic they carry, risk associated with suspension cable bridges is relatively high and requires good security risk management strategies. The complex nature of the suspension bridges, coupled with higher security threat and location, makes mitigating these threats costlier than for other bridges. This chapter discusses the security of suspension cable bridges in a risk management framework in such a way that it can be expanded to cover all aspects of risk management associated with suspension bridge upkeep.

It should be recognized that the security problem can only be resolved by finding an appropriate balance between hazard potential costs and mitigation costs. Risk, at its foundation, is built to address consequences, threats, and vulnerabilities all at once in a balanced and objective manner. Thus, risk management is the appropriate venue for addressing this issue. In order to achieve this, this chapter initially gives an overview of risk management basics as applied to security of suspension bridges. Five components that comprise risk management—assessment, acceptance, mitigation, monitoring, and communications—are explored. Then a large section of the chapter is devoted to implementing these risk components to suspension bridges. A three-tier assessment model that is specific to suspension cable bridges is then introduced. The model can be used as is by bridge owners or can be recalibrated to suit the bridge owners' assessment goals. This is followed by a discussion on general security mitigation measures that might be adopted by owners, and a brief exploration of the other three risk management components.

Two important emerging corollaries to risk management that need to be considered while managing bridge security are briefly discussed next:

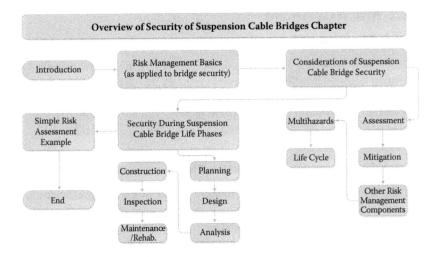

Overview of Security of Suspension Cable Bridges Chapter

Figure 12.1 Composition of this chapter.

multihazards and life cycle. Recognizing that bridge management operations during the life span of the bridge (planning, design, analysis, construction, inspection, and maintenance/rehabilitation) are intertwined with bridge security issues, near the end of the chapter a simplified overview of how to use these operations to improve bridge security is presented. The chapter ends with an illustrative example of security risk assessment of a suspension bridge. The main objective of this chapter content is to ensure that the suspension bridge owner or operator can provide for a secure bridge at reasonable costs. The composition of the chapter is shown in Figure 12.1.

12.2 RISK MANAGEMENT

12.2.1 Overview

This section provides an overview of risk management principles as applied to civil infrastructures in general and to suspension cable bridge security in particular. We follow Ettouney and Alampalli's (2012a) handling of risk management, as comprised of five components: assessment, acceptance, mitigation, monitoring, and communications. We explore each component in detail while providing, when pertinent, specific discussion on how the component might be dealt with in the field of suspension cable bridge security, and how various risk management components intersect with each other. We end this section with a brief comparison between the concepts of risk and resilience, as resilience applications are very important to ensure continuity of operations after major disasters. More information about bridge resilience can be found elsewhere (Alampalli and Ettouney, 2010; Ettouney and Alampalli, 2012b).

12.2.2 Risk Assessment

12.2.2.1 Overview

The first and most widely practiced component of risk management is the risk assessment component. Before we assess risk, its metrics have to be defined. There are several popular risk metrics that are used to address bridge security aspects. This section discusses these metrics and then summarizes some popular risk assessment methods used in the bridge security field.

12.2.2.2 Subjective versus Objective Risk

Defining risk scales/metrics is an essential step in ensuring efficient risk management to make sure that optimal decisions are made. These metrics can be subjective or objective. Subjective risk scales are described as ranks. For example, *high, medium,* and *low* is a three-step ranking that might describe risk (or any of its components). Five-step or even ten-step ranks have also been used in practice. Subjective risk scales are by definition all relative. They offer simplicity, yet they can't be used in demanding large analytical projects. Several authors have utilized this approach to describe risk in a subjective fashion (FEMA, 2005; DHS, 2011a, 2011b, 2011c).

Objective risk scales can be subdivided into two categories: relative or absolute. Relative objective risk or any of its components can be described objectively on an analytical scale, say, from 0 to 10, or 0 to 100. In most cases, lower scores indicate better situations. A perfect risk condition would be at the bottom of the scale, say, a risk of 0. The relative risk scales were also used by the Department of Homeland Security (DHS) (2011a, 2011b, 2011c). An absolute risk scale is generally measured with monetary units. An example would be a risk of US$50. The lower number indicates a better situation. Absolute risk scales have been used by many in the literature (Jorion (2007) in the field of finance and Ettouney and Alampalli (2012a, 2012b) in the field of civil infrastructure). A comparison between advantages and disadvantages of risk scales is shown in Table 12.1.

Table 12.1 Advantages and Disadvantages of Risk Scales

	Risk Scale	Advantages	Disadvantages
Subjective	Relative-discrete	Simple	Difficult to use for detailed risk management components other than assessment
Objective	Relative-continuous	Simple; can be used for large-scale assessment and project prioritizations	Requires more computational resources
	Absolute	Most useful of risk scale	Requires most computational resources; can be complex at times

12.2.2.3 Survey of Risk Assessment Methods/
Proposed Methodology

Risk is generically defined as an expression of a relationship between a particular hazard (or threat) that might degrade the performance of the infrastructure under consideration and the consequences that might result from such a degradation of performance (Gutteling and Wiegman, 1996; FEMA, 2005; NRC, 2010). Most communities (such as engineering, finance, insurance, and medicine) have adopted a variant of this particular definition of risk (Gutteling and Wiegman, 1996). For example, in the infrastructure security community, FEMA (2005) uses an objective risk definition that states

Risk rating = function (Consequences, Threat, Vulnerability)

$$= \text{function } (C, T, V)$$

Of course, the type of risk function depends on the desired degree of complexity of risk analysis. A simple risk rating definition was used by FEMA (2005):

$$Ri = C \cdot T \cdot V \tag{12.1}$$

A more involved methodology that utilized a form of Markov network was used by DHS (2011a, 2011b, 2011c). This method is based on utilizing affinities or interactions or links between risk variables (referred to as considerations in this chapter) in forming objective ratings for risk, Ri, consequences, C, threats, T, and vulnerabilities, V. Without accommodating such interactions/links, the resulting risk and its components might not be accurate (NRC, 2010). A detailed description of the Markov network and its use in risk analysis can be found elsewhere (Fenton and Neil, 2013; Koller and Friedman, 2009). Markov network risk modeling techniques are used for solving the bridge security example presented in Section 12.6. The Markov network method is fairly simple, yet it was shown to provide accurate results when applied to different types of civil infrastructures (DHS, 2011a, 2011b, 2011c).

12.2.3 Risk Acceptance

Setting acceptance thresholds is perhaps the most difficult step in asset (or community) management. Traditionally, a prescribed acceptance threshold has been adopted in civil infrastructure projects based on a reasonable, yet subjective, probabilistic nonexceedance value (Ettouney and Alampalli, 2016a, 2016b). Such a practice was reasonable since civil infrastructure projects were mostly based on ensuring safety. With the advent of risk-based paradigms, where project decisions are based on safety as well as cost (both life cycle costs and initial capital expenditures), setting risk acceptance thresholds became more difficult to define. Objectively, setting a reasonable acceptance threshold for security risk of the bridge under consideration is as important as defining/assessing security risk itself. For without setting a threshold, security improvement projects either might be unnecessarily costly or might result in lesser performance (Figure 12.2). Unfortunately, there is little known on objective methods that can be of help in setting reasonable security risk acceptance thresholds, but one can still make some general suggestions. A simple approach is to try to set the acceptance threshold so as to minimize total costs, as will be detailed next.

Even though it is not too obvious, security risk acceptance thresholds can have immense cost implications. The total costs of any event are composed of the total expected damage costs of the event itself and the costs of retrofits that might mitigate security risks. It is obvious that as costs of retrofits increase, security risks will decrease, and as a result, damage costs will decrease. The reverse is true. This relationship is shown in Figure 12.2. As shown in the figure, the total cost will have a "sweet spot" at which the total costs are minimum. If the security risk acceptance threshold moves away, in either direction, from this threshold, the total costs will increase— an undesirable result. Thus, decision makers should strive for a security risk acceptance threshold that is as close to this sweet spot threshold as possible. This can be achieved only by careful estimations of damage costs as well as retrofit costs.

There are two additional complications to this situation. An available budget is the first issue, as shown in Figure 12.3. If the available budget can only afford the decision maker a higher risk level than the sweet spot offer, then the resulting retrofits will be far from optimal. Similarly, if the target risk (desired risk rating) is much lower than either the sweet spot or the available budget, such a situation will result in nonoptimal risk situation, and it will also result in potential budget-security mismatch. Additional

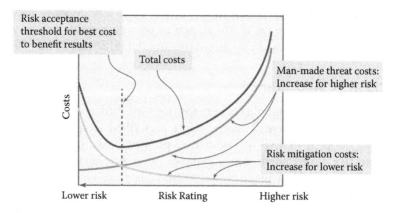

Figure 12.2 Optimal risk acceptance condition.

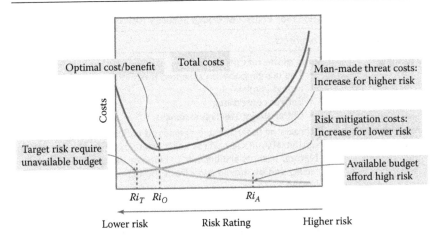

Figure 12.3 Practical risk acceptance conditions.

studies might be needed to see how an acceptable security risk versus available budget compromise might be reached.

Figure 12.2 shows only one level of a security-related damaging event. Similar illustrations can be made for all possible event levels. The security risk optimal acceptance limit is expected to differ for different event levels. Also, it is advisable to include two additional factors while looking for this security risk acceptance sweet spot: the life cycle costs and benefits and the effect of multihazards if the asset or communities are susceptible to more than one potentially damaging hazard.

12.2.4 Risk Mitigation

The risk mitigation is generally considered when risk rating of the bridge of interest is found to be higher than the risk acceptance threshold. The security risk mitigation component includes three phases:

I. Baseline mitigation
II. Project plans and prioritizations
III. Project execution

Phase I bridge security risk mitigation basically tries to decide upon the type of bridge security risk mitigation that is most efficient. There are four possibilities, as shown in Table 12.2.

If the outcome of Phase I is to embark on bridge security risk mitigation effort, the decision maker then moves on to Phase II, which is preparing potential risk mitigation projects. Bridge security risk mitigation projects should account for potential improvements to some or all of the CTV components. Many times, decisions are made to physically improve

Table 12.2 Decision Options for Phase I of Bridge Security Risk Mitigation/Treatments

Type of Action	Comments
Do nothing	Some of the reasons can be: • The margin between accepted threshold and actual risk ratings is minimal. • Budget constraints. • Other subjective circumstances.
Remove potentially damaging events	If threats or hazards can be removed entirely, it is a good option. Man-made hazards are a prime example for this option. Natural hazards are difficult to remove, except if the whole facility can be moved away from the exposure to the natural hazard.
Transfer responsibility	Reasons for potential interruption to operations might be transferred by opting for insurance. In case of many civil infrastructures, such as bridges or tunnels, it is normally not feasible.
Improve/treat deficiencies	Improvements include physical retrofits and operational improvements.

facilities that are usually accomplished by improving bridge vulnerability. Sometimes, it might be more efficient to improve threat or consequence ratings. Careful studies are needed to achieve a balance between the improvements of the CTV to achieve the optimal benefit–cost ratio. For example, it might be more efficient to improve access control or setback than bridge hardening. It is also very important to consider potential undesirable effects due to possible interactions between hazards before embarking on a bridge security rehabilitation effort.

After identifying potential candidate improvement projects, a project prioritization effort is needed to select the most suitable and cost-effective project to achieve the required risk threshold level. This should consider costs, benefits, project scale, and life cycle considerations (see Fenton and Neil (2013) for theoretical background and Ettouney and Alampalli (2012a, 2012b) for practical examples). The next section will include more details on Phase II security risk mitigation best practice considerations. After deciding on the most suitable project for risk improvements, Phase III entails project execution. Figure 12.4 shows the three phases of risk improvement.

12.2.5 Risk Monitoring

It is given that risk ratings of suspension bridges will tend to increase with passage of time as a result of potential increases in all three risk components: CTV. For example, physical deterioration (i.e., increased vulnerabilities) of bridges in general is well documented through deterioration rating studies (Agrawal and Kawaguchi, 2009). Mahmoud (2011) documented the deterioration issue of suspension cables as a function of time. The

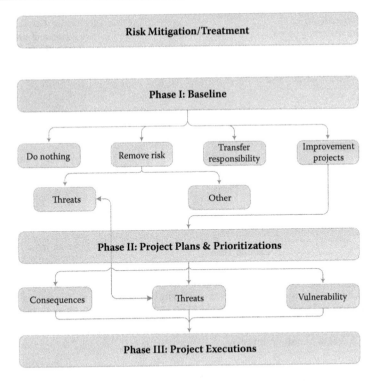

Figure 12.4 Three phases of bridge security risk mitigation/treatment.

rate of physical degradation can be steeper than the forecasted values, and the accepted threshold reached much faster than predicted by initial risk studies. Of course, the vulnerability component of risk is not the only risk component that increases with time. Security threats also tend to increase with time as target attractiveness or importance increases. Detailed discussions on target attractiveness can be found elsewhere (FEMA, 2005; DHS, 2011a, 2011b, 2011c). Finally, consequences do increase as time passes due to increased traffic volumes, associated cost of service interruptions, costs of physical replacements, and so forth. Due to this time dependency of risk and to avoid having a security risk rating above an acceptable threshold, the state of bridge risk should be monitored periodically as a component of a comprehensive risk management process, and appropriate modifications to the original plans should be made as needed to avoid the potential steep increase of the risk of a security-related event based on the plans. The value of risk monitoring is shown in Figure 12.5. Risk-monitoring techniques should cover all three components of risk (C, T, and V). Some of the general risk-monitoring techniques are shown in Table 12.3.

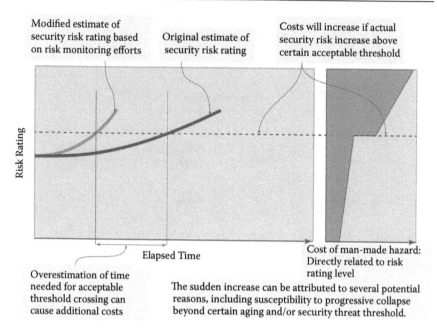

Figure 12.5 Value of risk monitoring.

Table 12.3 Examples of Risk-Monitoring Techniques

Component	Monitoring Techniques
Consequences	• Average daily traffic (ADT) • Weigh-in-motion data for demands • Costs of service interruptions • Costs of physical retrofits and replacements • Changes in other potential consequences (see DHS, 2011a, 2011b, 2011c)
Threats	• Security monitoring techniques (see FEMA, 2005) • Follow changes in target attractiveness issues (see DHS, 2011a, 2011b, 2011c)
Vulnerabilities	• Structural health monitoring (SHM) and nondestructive testing (NDT) methods (see Ettouney and Alampalli, 2012a, 2012b) • Pertinent inspection and maintenance processes (see Table 12.4)

12.2.6 Risk Communications

The final and perhaps most important component in risk management is risk communications. Communications is the bridge between the technical and professional community, decision makers, elected officials, funding sources, and the public at large. There are numerous objectives to risk communications, as shown in Table 12.5. Obviously, all of the above objectives are important. Unfortunately, risk communications usually receives

Table 12.4 Vulnerability Monitoring/Inspection Practices and Security Issues

Type of Inspection/ Monitoring	Description	Impact on Security Consideration
Routine	Visual inspection	Can also result in some observations that have security implications
Internal	In-depth testing of cables	By detecting degradations in properties, as-is robustness of cables to resist blast can be estimated rather than the design robustness
SHM	Modal and parameter testing and global behavior of the bridge	Potential intersections/interactions between hazards can be estimated
NDT	Local area or specific component testing	By detecting degradations in properties, as-is robustness of cables to resist blast can be estimated, rather than the design robustness

Table 12.5 Objectives of Risk Communications

Objectives	Comments
Education	This includes educating decision makers (message can be technical or nontechnical) and the public (message would be nontechnical).
Security warnings	Communicating upcoming hazard or threat and susceptibility of community or asset.
Behavior change	Communicating status of some of the risk components, and aiming to modify the behavior of stakeholders to improve or comply with different needs to improve risk. The considerations of Tier III of Tables 12.9 through Table 12.22 are prime candidates for behavioral change topics that relate to suspension cable bridge security.
Mitigation efforts	Informing bridge stakeholders (public, bridge managers, etc.) of upcoming or ongoing bridge security improvement efforts as needed.
Funding	Relaying needs for funding to elected officials, as well as federal, state, and local agencies.

the least degree of attention, whereas, due to its paramount importance, it should receive the most attention. Lasswell (1948) and Gutteling and Wiegman (1996) subdivided risk communications into five essential modules: originator (source), content (message), recipient, medium (channel), and objectives (destination). The five modules are interrelated as shown in Figure 12.6. Any successful risk communication plan has to accommodate all five modules in a careful and well-thought-out manner. Additional discussions of risk communications for civil infrastructures in general can be found in Ettouney and Alampalli (2016a, 2016b).

12.2.7 Risk versus Resilience

Resilience is an emerging concept that aims at ensuring continuity of operations and functions at an acceptable rate during and after any man-made

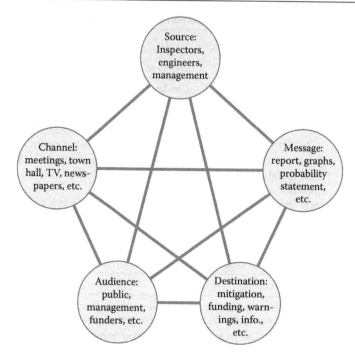

Figure 12.6 Risk communication modules as pertaining to suspension bridges.

or natural disaster (e.g., see NIAC, 2009). Ettouney and Alampalli (2012a) argued that resilience can be considered a special case of risk. This is due to the fact that resilience has four main components: robustness, resourcefulness, recovery, and redundancy (NIAC, 2009). These components are sometimes referred to as the 4Rs. We can argue that the 4Rs are interrelated to the three risk components (consequences, threats, and vulnerabilities), as shown in Table 12.6.

Thus, resilience might be recast as a special form of risk. It should be noted that resilience focuses mainly on continuity of operations and the degree and length of operational interruptions. Thus, if we cast risk consequences, the threat and vulnerability parameters that affect these two types of consequences exactly define the resilience. The interrelationship between risk and resilience is shown in Table 12.7 and Figure 12.7. Table 12.7 shows

Table 12.6 Relationships between Risk and Resilience

Risk Components	Resilience Components			
	Robustness	Resourcefulness	Recovery	Redundancy
Consequences	Minor	Major	Major	Major
Threat	Major	Minor	Minor	Major
Vulnerability	Major	Minor	Minor	Major

Table 12.7 Comparison between Risk and Resilience

Item	Risk	Resilience
Components	• Vulnerability • Threat • Consequences	• Robustness • Resourcefulness • Recovery • Redundancy
Management components	Assessment, acceptance, treatment/improvement, monitoring, communications	
Metric properties	• Objective • Subjective • Relative • Absolute	
Main emphasis	Cost or cost benefits	Continuity of operations

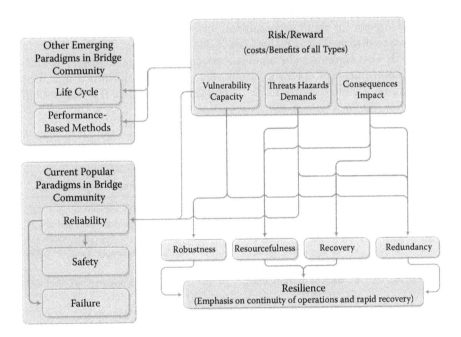

Figure 12.7 Risk and resilience interrelationship.

comparisons of different management issues that concern risk and resilience. Figure 12.7 shows how risk is a superset of resilience. It also shows how risk can be considered a superset of reliability, safety, failure, life cycle analysis, and performance-based methods, which are of interest to the bridge community. A detailed discussion of these interrelationships can be found elsewhere (Ettouney and Alampalli, 2012a, 2012b, 2016a).

12.3 ELEMENTS OF SUSPENSION BRIDGE SECURITY

12.3.1 Overview

Risk-based management for handling security of civil infrastructures such as cable suspension bridges has been recommended by DHS (2009). Thus we offer an example of risk assessment procedure for this type of bridges in the remainder of this chapter. Objective bridge security risk management effort requires a top-down approach. Thus, at first, management efforts have to be concentrated on the three risk components, C, T, and V, and then build on these efforts. This can be accomplished by subdividing the bridge security issue into three tiers to make it easier to objectively manage the security issue in a risk management framework.

Tier I: This tier can compute all risk management components easily and also can interact with reliability (safety) concerns since they are of interest for estimating vulnerabilities and threats. Tier I ratings are inferred from Tier II ratings (see Table 12.8).

Tier II: This comprises the main components for bridge security. For illustration purposes, 14 main components have been identified and their ratings are inferred from Tier III ratings. For each Tier II consideration/

Table 12.8 Cable Bridge Security for Tiers I and II Considerations

No. Tier II	Tier I			Tier II Considerations	Is This Suspension Bridge Cable Specific?	Type of Required Analysis
	C	T	V			
1	Y	Y	Y	Site layout	N	Traffic/WIM/architectural/social
2	Y	Y	Y	Bridge component (suspension cable)	Y	Structural/inspection/SHM/NDT
3	Y	Y	Y	Bridge components (suspenders)	Y	Structural/inspection/SHM/NDT
4	N	Y	Y	Bridge component (other)	N	Structural/inspection/SHM/NDT
5	Y	N	Y	Bridge system	N	Structural/inspection
6	Y	Y	Y	Nonstructural	N	Structural/inspection
7	N	N	Y	Access control	N	Security
8	N	N	Y	Deterrent and other security measures	N	Security
9	Y	Y		Use/functional	N	Functional
10	Y	N	Y	Operational	N	Management
11	Y	N	Y	Special needs	Y	Structural/inspection/SHM/NDT
12	Y	N	N	Economics/social	N	Social/economic
13	N	Y	Y	Other demands	Y	Monitoring
14	N	Y	N	Additional subjective threat factors	N	Social/economic/functional

component, from a security view point, evaluation of all relevant Tier III components under it, vulnerabilities of these components, acceptance/ mitigation methods, how to inspect/maintain/rehab, and what to avoid is needed. Care must be exercised not to implement a mitigation effort for one hazard by creating vulnerability of the component of another hazard.

Tier III: This comprises the most detailed components of cable bridge security. Sixty-six such components have been identified in this chapter for illustration purposes. The remainder of this section offers a detailed presentation on assessing these three tiers.

12.3.2 Assessment

This section identifies Tier III issues for bridge security risk evaluation. Each Tier II component, along with associated Tier III components, is presented in Tables 12.9 through 12.22. This information is neither complete nor limited, and thus can be shortened or expanded as needed by the bridge owner. The risk assessment efforts should evaluate each of these Tier III considerations. Evaluation could be subjective or objective and probabilistic or deterministic, depending on the amount of effort, detail type, and methodology used.

12.3.3 Mitigation

It was argued earlier that an efficient bridge security risk mitigation effort will invariably include three phases (Figure 12.4). Table 12.23 shows some basic insights on the best practices that might be followed when considering the projects of Phase II for different Tier II bridge risk considerations.

12.3.4 Risk Acceptance, Monitoring, and Communication

We discussed earlier the three risk acceptance limit states (Figure 12.3), which include the sweet spot that gives the optimal balance between security level and project costs, budget limitation situation, and selection of target risk when budget is not an issue. Although it is difficult, some practical

Table 12.9 Tiers II and III Issues for Site Layout

Tier II (Inferred)	Number	Tier III (Observable/Computed)	Comments
I—Site layout	1.01	Average daily traffic (ADT)	From WIM or other sources
	1.02	Locality (urban, suburban, etc.)	See DHS (2011b, 2011c)
	1.03	Adjacent infrastructure (including buildings)	See DHS (2011b, 2011c)
	1.04	Setback rating (for site in general)	See FEMA (2005) and DHS (2011a, 2011b, and 2011c)

Table 12.10 Tiers II and III Issues for Suspension Cable System

Tier II (Inferred)	Number	Tier III (Observable/Computed)	Comments
2—Bridge component (suspension cable)	2.01	Number of independent suspension ropes or cables; see Figure 12.8	Bridge inventory; might require advanced analysis
	2.02	Number of suspension ropes that can be damaged without causing progressive collapse	Bridge inventory; might require advanced analysis
	2.03	Condition rating(s)	From inspection reports; see Figure 12.9
	2.04	Control (minimum) setback	Specific to suspension cables; see FEMA (2005) and DHS (2011a, 2011b, 2011c)
	2.05	Multihazard index	Might require analysis; see Ettouney and Alampalli (2016b) and Ettouney et al. (2005)
	2.06	Blast retrofit	Blast retrofit can improve vulnerability
	2.07	Accumulated damage	Might require advanced inspection or analysis; see Mahmoud (2011) and FHWA (2012)
	2.08	Estimated remaining life	Might require advanced inspection and analysis. See Mahmoud (2011) and FHWA (2012)

steps can be taken to arrive at a reasonable risk acceptance threshold that balances budget and security level. These include ensuring that curves similar to those of Figure 12.3 are produced in an objective manner while accounting for multihazard interactions, life cycle effects, and uncertainty nonexceedance levels that are consistent with state of the practice. Note that life cycle costs have a tendency to improve benefits outlook, thus reducing budget gaps, if present.

Risk monitoring is an emerging field that includes numerous knowledge gaps. A good practice is to make an attempt to monitor the most important considerations of Tiers II and III. There is a tendency to monitor vulnerability-related issues at the expense of monitoring threat- or consequence-related issues. Doing so will lead to inaccurate risk rating. A balanced monitoring effort should include processes to monitor all risk components evenly. Monitoring and archiving those considerations to estimate risk ratings in (near) real time is not an easy task. However, the value of monitoring, as discussed earlier in this chapter, might justify the effort.

Figure 12.8 Multiple suspension cables offer a higher degree of redundancy. (Courtesy of Weidlinger Associates, Inc., New York, NY.)

Figure 12.9 Inspection of suspension cables. (Courtesy of William Moreau.)

Table 12.11 Tiers II and III Issues for Suspender System

Tier II (Inferred)	Number	Tier III (Observable/Computed)	Comments
3—Bridge components (suspenders)	3.01	Number of suspension ropes; see Figures 12.10 and 12.11	Bridge inventory; might require advanced analysis
	3.02	Number of suspenders that can be broken in a series	Bridge inventory; might require advanced analysis
	3.03	Condition rating	Inspection reports
	3.04	Control (minimum setback)	Specific to suspenders; see FEMA (2005) and DHS (2011a, 2011b, 2011c)
	3.05	Multihazard index	Might require analysis; see Ettouney and Alampalli (2016b) and Ettouney, et al. (2005)
	3.06	Blast retrofit	Blast retrofit can improve vulnerability
	3.07	Accumulated damage	Might require advanced inspection and analysis; see Mahmoud (2011) and FHWA (2012)
	3.08	Estimated remaining life	Might require advanced inspection and analysis; see Mahmoud (2011) and FHWA (2012)

Figure 12.10 Suspenders and suspension cables. (Courtesy of William Moreau.)

Figure 12.11 Connection of a bridge suspender in good condition. (Courtesy of Weidlinger Associates, Inc., New York, NY.)

Table 12.12 Tiers II and III Issues for Other Bridge Components

Tier II (Inferred)	Number	Tier III (Observable/Computed)	Comments
4—Bridge component (other)	4.01	Vessel collision (pier type, pier protection, VC under [vertical clearance under])	Might require advanced analysis
	4.02	Vehicle collision (VC above [vertical clearance above], distance between suspenders/truss and leftmost or rightmost lane)	Might require advanced analysis
	4.03	Bearing type, setback, and condition	Subjective estimate; might require advanced analysis

Table 12.13 Tiers II and III Issues for Bridge System

Tier II (Inferred)	Number	Tier III (Observable/Computed)	Comments
5—Bridge system	5.01	Span type	Subjective estimate
	5.02	Stiffening truss is available or not; see Figure 12.13	Subjective estimate; might require advanced analysis
	5.03	Fracture critical or not	Subjective estimate; might require advanced analysis
	5.04	Fire protection	Subjective estimate; might require advanced analysis in case of fire susceptibility
	5.05	Multihazard analysis performed during design	Requires analysis; see Ettouney and Alampalli (2016b) and Ettouney et al. (2005)
	5.06	Level of redundancy available	Subjective estimate; might require advanced analysis to accurately estimate redundancy

(continued)

Table 12.13 (continued) Tiers II and III Issues for Bridge System

Tier II (Inferred)	Number	Tier III (Observable/Computed)	Comments
	5.07	Progressive collapse considerations	Can be estimated via analysis
	5.08	Deck type (concrete, orthotropic, FRP [fiber reinforced polymers], etc.)	Subjective estimate; might require advanced analysis to accurately estimate deck robustness
	5.09	Approach types and characteristics; see Figure 12.12	Subjective estimate; might require traffic and robustness analysis
	5.10	Load restrictions	Load restrictions indicate deficiency of robustness for live loads; this is an indication of vulnerability to blast loads too; special care and advanced analysis are needed to accommodate such a condition
	5.11	Height postings	Height posting might have an effect on blast threat rating; special analysis is required to estimate such an effect
	5.12	Year built (quality of material and construction are controlled by this)	Older bridges can have degraded performance, which might increase vulnerability
	5.13	Year of last retrofit (quality of material and construction are controlled by this)	Retrofits can improve performance
	5.14	Number of bridges separated vertically; see Figure 12.14	Subjective estimate; might require advanced analysis
	5.15	Ramps underneath	Subjective estimate; might require advanced analysis

Risk communication is very important aspect of risk management. Good communication plans should be clear and consistent with the intended audience, accurate and never misleading, uniform and consistent with how other hazards are communicated with relevant stakeholders, complete without divulging any sensitive information, and have a good balance between objectivity and subjectivity. The plan might resort to comparisons if pertinent and should not raise any doubts if subjective. The communication has to be balanced between confidentiality needs and public awareness. Communication plans should be periodically checked for effectiveness via feedback from the stakeholders and revised as needed based on the input received. A good and complete communication plan should accommodate all five components of Figure 12.7.

Figure 12.12 Although this type of suspension bridge approach is no longer used, its mass and setback offer a good security protection. (Courtesy of Weidlinger Associates, Inc., New York, NY.)

Figure 12.13 Typical stiffening trusses in suspension bridges. (Courtesy of William Moreau.)

Figure 12.14 Multilevel suspension bridge. (Courtesy of Weidlinger Associates, Inc., New York, NY.)

Table 12.14 Tiers II and III Issues for Nonstructural Bridge Systems

Tier II (Inferred)	Number	Tier III (Observable/Computed)	Comments
6— Nonstructural	6.01	Drainage	Might have a multihazard effects during storms/ floods
	6.02	Debris accumulation	Might have a multihazard effects during storms/ floods
	6.03	Joint/deck condition	Might have multihazard effects
	6.04	Lifeline use (gas, water, and or electric conduits); see Figure 12.15	Can have effects on consequences when computing risk ratings

Table 12.15 Tiers II and III Issues for Access Control

Tier II (Inferred)	Number	Tier III (Observable/Computed)	Comments
7—Access control	7.01	Physical barriers	Can affect overall or specific setbacks
	7.02	Access to towers; see Figure 12.16	Access needs to be limited to authorized personnel only
	7.03	Access to climb suspension bridge ropes	Access needs to be limited to authorized personnel only
	7.04	Access to anchorage areas; see Figure 12.17	Access needs to be limited to authorized personnel only
	7.05	Access for emergency personnel such as fire department	Access needs to be limited to authorized personnel only

Figure 12.15 Lifeline conduits on a suspension bridge cable. (Courtesy of Weidlinger Associates, Inc., New York, NY.)

Figure 12.16 Typical tower system in a suspension bridge. (Courtesy of William Moreau.)

Figure 12.17 Anchorage area in a cable bridge. (Courtesy of William Moreau.)

Table 12.16 Tiers II and III Issues for Deterrence and Other Security Measures

Tier II (Inferred)	Number	Tier III (Observable/Computed)	Comments
8—Deterrent and other security measures	8.01	Security cameras; see Figure 12.18	Well-placed security cameras can have deterrence effects
	8.02	Presence of police	A very efficient security measure
	8.03	Other	See DHS (2011a, 2011b, 2011c) and FEMA (2005)

Figure 12.18 Security camera on a suspension bridge. (Courtesy of William Moreau.)

Table 12.17 Tiers II and III Issues for Use and Functional Considerations

Tier II (Inferred)	Number	Tier III (Observable/Computed)	Comments
9—Use/ functional	9.01	Pedestrians	See DHS (2011a, 2011b, 2011c) and FEMA (2005)
	9.02	Trains	See DHS (2011b, 2011c)
	9.03	Vehicles	See DHS (2011b, 2011c)
	9.04	Combined/shared use	See DHS (2011b, 2011c)
	9.05	Functional use	See NYSDOT vulnerability manuals

Table 12.18 Tiers II and III Issues for Operational Considerations

Tier II (Inferred)	Number	Tier III (Observable/Computed)	Comments
10— Operational	10.01	Scenario analysis and drills	See DHS (2011b, 2011c)
	10.02	Inspection aspects (SHM, NDT, visual only)	See Ettouney and Alampalli (2012b)
	10.03	Inspection frequency	Increased inspection frequency can detect degradation of security measures early on

Table 12.19 Tiers II and III Issues for Special Needs of Suspension Bridges

Tier II (Inferred)	Number	Tier III (Observable/Computed)	Comments
11—Special needs	11.01	In-depth cable inspection	Special inspection can detect degradation of cables early on; see Figure 12.9

Figure 12.19 Maintenance of cable bridge underway. (Courtesy of William Moreau.)

Table 12.20 Tiers II and III Issues for Economic/Social Considerations

Tier II (Inferred)	Number	Tier III (Observable/Computed)	Comments
12—Economics/social	12.01	Downtime after disaster	Effective resilience plans should be prepared to ensure short downtime after disasters
	12.02	Replacement value	This can have a major effect on consequences in a risk-based management operation; see DHS (2011b, 2011c)
	12.03	Network resiliency	Bridge network needs to be accommodated in any effective resilience plans to ensure short downtime after disasters; this will limit undesired consequences after any disaster

Table 12.21 Tiers II and III Issues for Additional Demands

Tier II (Inferred)	Number	Tier III (Observable/Computed)	Comments
13—Other demands	13.01	Wind strength	Wind can play a major role in the vulnerability rating for cable bridges in real time
	13.02	Temperature variations	Extreme temperature swings can play a major role in the vulnerability rating for bridges

Table 12.22 Tiers II and III Issues for Additional Threat/Consequence Factors

Tier II (Inferred)	Number	Tier III (Observable/Computed)	Comments
14—Additional subjective threat factors	14.01	Significance of bridge	This can have a major effect on threat and consequences in a risk-based management operation; see DHS (2011b, 2011c)
	14.02	Function criticality	This can have a major effect on threat and consequences in a risk-based management operation; see DHS (2011b, 2011c)

Table 12.23 Suggested Basic Mitigation Options

Tier II Issue	Mitigation Options
Site layout	Fencing Relocation of walkways, trails, etc. Drainage Scour
Bridge components (suspension cables)	Access control Surveillance Blast protection cell Fireproofing Remove/reduce encroachment due to oversize vehicles More cables (redundancy)
Bridge components (suspenders)	Armoring Redundancy Surveillance Stiffening truss spanning more than one bay Fireproofing
Bridge components (other)	Lighting Surveillance
Bridge system	Surveillance Lighting Wider lanes Vertical clearance for boats SHM
Nonstructural	Adequate drainage Setback
Access control	Location of toll booths farther away from bridge Setback Pedestrian fencing Location of sidewalks Locked gates/doors Restricted access to anchorage area, towers, etc. Security checks and IDs for inspectors and others involved
Deterrence and other security measures	Cameras Fencing Physical security on site SHM Intrusion alarms
Use/functional	Restricting trucks Restricting width of vehicles
Operational	Site distance Cameras for maintenance SHM Debris accumulation control Design of ramps for smooth flow and avoiding accidents
Economy/social	Not allowing recreational access/tours Restricting pedestrian access
Other demands	Suicide fencing Minimizing wind loads
Additional subjective threats/consequences	Suicides

12.4 OTHER SECURITY-RELATED CONSIDERATIONS

12.4.1 Security and Multihazards

Security (bomb/blast hazard), which is the focus of this chapter, does interact with other natural hazards that affect cable-supported bridges to various degrees. These interactions can be beneficial or complementary; that is, they can improve the overall structural performance. For example, improving fatigue life generally improves structural performance to blast pressures, wind, or live loads. The interactions can also be contradictory; that is, they can reduce the overall structural performance. For example, adding mass (hardening) can improve blast response, but would degrade seismic response, if not designed carefully. These interactions form the basis for the theory of multihazards introduced by Ettouney et al. (2005). A corollary to the theory is that any decision made regarding any one hazard will have an effect on how the structure will respond to any other hazard (Ettouney and Alampalli, 2012a, 2012b).

The multihazard interaction matrix (MHIM) is an objective way to assess the interaction of hazards (see Alampalli and Ettouney, 2008; DHS, 2011a, 2011b, 2011c). A cell in the ith row and jth column in the matrix quantifies the effects of a decision made regarding the ith hazard on the response of the structure due to the jth. Among the properties of the MHIM are that data in the matrix are expressed in percent in general, the matrix is not necessarily symmetric, and the main diagonal is always 100%. Even though the theory is simple in nature, developing an MHIM is not an easy task. The intricacies of how a bridge responds to different hazards during its service life, as well as the decisions resulting from observing these responses, need to be studied carefully. This is more important for cable-supported bridges due to their size, volume of traffic they carry, and costs associated with their maintenance. Alampalli and Ettouney (2013) explored several implications of accommodating the interactions of security hazards with other hazards for cable-supported bridges. They showed that ignoring multihazard interactions can lead to an undesirable safety/security outcome for those types of infrastructures. A detailed study of theoretical and practical issues regarding multihazard considerations for civil infrastructures can be found elsewhere (Ettouney and Alampalli, 2016b).

12.4.2 Life Cycle and Security Analysis

Life cycle analysis (LCA) should be an integral component of any risk management process. Ettouney and Alampalli (2012a, 2012b) argued that there are three parts of LCA: costs, benefits, and life span. For a proper LCA, the decision maker needs to account for time-based changes in costs and benefits over the life span of interest. One of the most common shortcomings while embarking on security improvement projects is that decision makers fail to consider LCA during their project planning. This often results in unnecessarily costly projects that fail to safeguard the desired security level. Some ways to accommodate proper LCA during project planning follow:

- Consider LCA for all risk management components: assessment, acceptance, mitigation, monitoring, and communications.
    - Assessment: Allow for reasonable physical deterioration and future operational changes.
    - Acceptance: Allow for all changes in all risk components—consequences, threats, and vulnerabilities.
    - Mitigation: Consider all current and future aspects of the projects, including interactions between all internal and external considerations.
    - Monitoring: Project all monitoring plans over an objectively computed life span.
    - Communications: Perhaps the most essential and least practiced component in an LCA effort. Remember that all stakeholders need to buy in to an LCA-based project. As such, the communication plans need to be developed with a high degree of efficiency.
- Define the life span objectively before embarking on any LCA effort. Conduct the objective analysis of life span on realistic assumptions.
- Defining life cycle costs (LCCs), although difficult, is much easier than defining life cycle benefits (LCBs) in an objective manner. Because of this, many owners ignore the objective definitions of LCBs. It is recommended that LCBs be defined with the same accuracy as LCC.
- Consider during the life span all potential interactions with other hazards. This is perhaps the single most important LCA step that is neglected by planners. It is also the step where most cost savings can be realized.
- Do not ignore uncertainties during LCA efforts.

12.5 SECURITY WITHIN THE BRIDGE LIFE SPAN

We identified numerous suspension bridge security considerations in Section 12.3. Now we turn our attention to how security considerations intersect with typical phases of the bridge life span. There are six typical distinct phases in bridge life span: planning, design, analysis, construction, inspection, and maintenance/rehabilitation. Implementing security considerations for new bridges is fairly different from implementing security measures for existing bridges. However, the principles are almost the same, and thus no distinction is made between new and existing bridges in this section.

12.5.1 Planning

Numerous security considerations can be implemented during the bridge planning stage. Some of these planning stage considerations are shown in Table 12.24.

Table 12.24 Security-Related Bridge Planning Efforts

Section 12.3 Consideration Reference	Description of Security Measures
1.01–1.04	Site planning issues (mainly for new bridges)
2.04	Setback for suspension cables
3.04	Setback for suspenders
4.03	Setback for components
5.01–5.15	Security of different aspects of a bridge system can be improved/controlled
6.01–6.04	Security of most nonstructural issues can be controlled with adequate planning
7.01–7.05	Access control issues can be addressed
8.01–8.03	Physical security and other deterrence systems can be improved by careful planning
9.01–9.05	Use and functional aspects of the bridge can be improved
14.01, 14.02	Target attractiveness issues can be reduced during planning

12.5.2 Design

Design is a bridge phase that occurs throughout the bridge life span due to possible rehabilitation of various components of suspension bridges because of their size, longer life span, and higher demands. Thus, with careful design plans, security considerations can be implemented at almost any stage of the bridge life. Table 12.25 shows some security considerations that can be implemented during the bridge design stage.

12.5.3 Analysis

As stated earlier in Section 12.3.2, analysis might be needed while assessing certain considerations of bridge security. Table 12.26 shows some of these considerations where analysis might prove to be of importance for assessment of bridge security.

Table 12.25 Security-Related Bridge Design Efforts

Section 12.3 Consideration Reference	Description of Security Measures
2.01, 2.02, 2.05, 2.06	Different aspects of suspension cable designs
3.01, 3.02, 3.05, 3.06	Different aspects of suspender designs
4.01–4.03	Vehicle and vessel collision issues, bearing design issues
5.01–5.15	Different aspects of bridge system design
6.01–6.04	Lifeline designs, deck design, other
7.01–7.05	Different aspects of access control
13.01, 13.02	Different loading demands, e.g., wind, temperature, earthquakes, etc.

Table 12.26 Security-Related Bridge Analysis Efforts

Section 12.3 Consideration Reference	Description of Security Measures
2.01, 2.02, 2.05, 2.07, 2.08	Analysis of different aspects of suspension cables
3.01, 3.02, 3.05, 3.07, 3.08	Analysis of different aspects of suspenders
4.01–4.03	Vehicle and vessel collision, different aspects of security of bearings
5.01–5.15	Bridge system analysis aspects
13.01, 13.02	Wind, temperature, and other loading demands

12.5.4 Construction

Construction of new bridges offers a chance to implement several security measures, as shown in Table 12.27.

12.5.5 Inspection

Inspection can offer a great contribution to several basic bridge security measures. Table 12.28 offers some instances where bridge inspection can provide assistance to security measures. More details on inspection processes for suspension bridges can be found in NCHRP (2004).

Table 12.27 Security-Related Bridge Construction Efforts

Section 12.3 Consideration Reference	Description of Security Measures
5.12	Year built (quality material and construction are controlled by this)
5.13	Year of last retrofit (quality material and construction are controlled by this)
10.02	Inspection aspects (SHM, NDT, visual only)

Table 12.28 Security-Related Bridge Inspection Efforts

Section 12.3 Consideration Reference	Description of Security Measures
2.03, 3.03	Condition rating can be used to determine vulnerabilities and need for mitigation
2.08, 3.08	Estimated remaining life can be used to determine vulnerabilities and need for mitigation
5.10, 5.11	Load or height postings
10.02, 10.03, 11.01	Inspection aspects (SHM, NDT, visual only), inspection frequency, special inspections

12.5.6 Maintenance and Rehabilitation

Bridge maintenance (Figure 12.19) is invaluable for the well-being of the bridge during its normal operating life. We discuss in this section how even some of the fairly basic maintenance works can also enhance in many ways the security of the bridge.

12.5.6.1 Recurring Major Works and Security

Recurring major works, on first look, do not appear to intersect with the security needs of cable-supported bridges. However, upon further reflection, there are great deal of intersections that require careful decision-making processes. We divide those intersecting needs into direct (immediate) and indirect (long-term) categories. The long-term effects will be mostly environmental, which will also lead to long-term/life cycle costs. Both are of consequence to security decisions and cost implications (initial and life cycle costs). Tables 12.29 and 12.30 summarize both categories.

12.5.6.2 Less Frequent Maintenance and Security

Less frequent maintenance results from normal wear and tear, responses to less frequent hazards (collision, earthquakes, high wind, etc.), long-term degradations (e.g., corrosion), increased traffic demands, and other effects. Security needs intersect with some of these different maintenance efforts, as shown in Table 12.31.

12.5.6.3 Other Security Enhancement Maintenance Efforts

Table 12.32 shows some security considerations that can be enhanced by maintenance efforts.

12.6 SECURITY RISK MODEL

This section presents an illustrative example on a method to exercise cable bridge security risk management components discussed earlier in this chapter. Earlier in this chapter, we identified 14 Tier II security components and 66 Tier III components. Any objective risk management process should include most or all these Tier III components in one form or another. In practice, risk models would include many more considerations than these issues. For example, the DHS (2011a) risk model contained more than 250 considerations. This section, however, presents a reduced-risk model to illustrate most of the basic functional components of the bridge security risk management process with the understanding that for an accurate risk management process, much more resolution and details are still needed, beyond what is offered in this section.

Table 12.29 Direct Effects

| Recurring Major Work | Physical | | | Level of Security Concerns | | Operational | | |
	Strength	Chemical, Biological, and Radiological (CBR) Release	Surveillance	Crowding (traffic)	Interruption of Normal Operations	Arson	Vandalism
Coating/painting	Low	Low	Medium	Medium	Medium	Medium	Low
Sealers	Low	Low	Medium	Medium	Medium	Low	Medium
Redecking	Medium (during construction)	Low	Medium	High	High	Low	Medium

Table 12.30 Long-Term Effects (Sustainability)

Qualitative Interaction between Security and Sustainability Concerns		Sustainability		
		Coating/Painting	Sealers	Redecking
Security	Coating/painting	Medium	Medium/low	Medium
	Sealers	Medium/low	Medium	Medium
	Redecking	Medium	Medium	Medium

Table 12.31 Intersections of Security and Less Frequent Maintenance Efforts

	Maintenance Effort		Security Consideration	
Component	Reason for Needed Maintenance/ Rehabilitation Effort	Potential Type of Action	Impact on Security Consideration	Reasoning
Stiffening girders/trusses Cables or suspenders Main towers Anchorages Saddles	Traffic increase, damage, deterioration/ corrosion/ aging	Hardening, replacing	High	By changing mechanical properties (mass, stiffness, etc.), the design bases' robustness of the component will change, thus affecting security level

Table 12.32 Security-Related Bridge Maintenance Efforts

Section 12.3 Consideration Reference	Description of Security Measures
2.05, 3.05	Multihazard index
5.12, 5.13	Age, year of last retrofit (quality material and construction are controlled by this)
8.01–8.03	Security measures and equipment
10.02	Inspection aspects (SHM, NDT, visual only)
12.01–12.03	Operational resilience and improvements

12.6.1 Risk Model

As discussed above, the complete cable bridge security risk process is demanding and requires a large effort, such that we can't exercise it herein. In order to give a complete practical example of risk-based assessment and mitigation decision making, we will use a reduced model that illustrates most of the major components (sans components such as risk acceptance, risk monitoring, risk communication, and multihazard considerations) of bridge security risk management efforts. The Bayesian network method is used due to the advantages it offers: it has the ability to represent the uncertainty foundations of the risk phenomena, input to the method can be subjective or objective without any loss of generality, all interrelationships and interactions between all variables of the problem can be accommodated in a relatively simple manner, it lends itself naturally to objective decision-making processes, it has a simple and straightforward capability, and it has the ability to communicate in a simple manner to all stakeholders due to its graph theory foundations.

Figure 12.20 shows the reduced cable bridge security risk model in a Bayesian network graph display. Note that the model contains all Tier I risk components (C, T, and V), and most of the Tier II components. The model shows the interrelationships between the risk variables. It also shows how the observation variables (defined as the variables that can be observed and estimated by the users of the network, such as site and physical vulnerability) flow into and produce the required risk information.

12.6.2 Variables/Nodes

The reduced-risk model includes 10 variables (risk network nodes). Each of the variables was described earlier in this chapter. The basic assumption is that any risk network analysis admits that all of its underlying variables are random variables with varying degrees of uncertainty. The next step is to decide upon a representation of these random variables. In this current model, for simplicity, it was assumed that all of variables are discrete with a small number of ranks that describe the state of each variable. Table 12.33

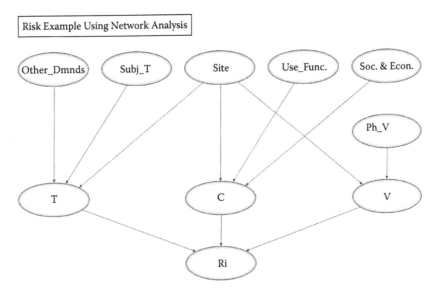

Figure 12.20 Simplified security risk model.

Table 12.33 Network Variables, Ranks, and States

Variable	Rank	States
Site	2	High, low
Physical vulnerability	3	High, medium, low
Vulnerability	3	High, medium, low
Use function	2	High, low
Other demands	2	High, low
Subjective threats	2	High, low
Social and economic	3	High, medium, low
Consequences	3	High, medium, low
Threat	3	High, medium, low
Risk	3	High, medium, low

shows the variables, their ranks (number of discrete state variables represented), and the possible states of each variable.

12.6.3 Conditional Probabilities Tables

The next component in building a network risk model is establishing the conditional probabilities tables (CPTs). A detailed description of CPTs can be found
elsewhere (Fenton and Neil, 2013; Koller and Friedman, 2009; Neapolitan,
2004; Ettouney and Alampalli, 2016a). CPTs can be established using several
methods, such as historical data, objective judgment, and personal expertise.
It is essential to note that the accuracy of the CPTs can only be assured by vali-
dating the model in which they are used (Fenton and Neil, 2013; DHS, 2011a,
2011b, 2011c; Vose, 2009; Ettouney and Alampalli, 2016a). Tables 12.34
through 12.43 show the CPTs that are used in conjunction with the risk model
shown in Figure 12.20. Note that each table includes the histogram of the
used conditional probability (CP) of interest. These CPs/histograms are generated only for illustrative purposes and need modification to accommodate
individual situations based on relevant experience and judgment.

Table 12.34 Site CPT

Site	High	0.7082
	Low	0.2918

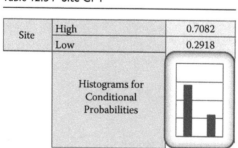

Histograms for
Conditional
Probabilities

Table 12.35 Physical Vulnerability CPT

Ph_V	High	0.5129
	Med.	0.3577
	Low.	0.1294

| Histograms for Conditional Probabilities | |

Table 12.36 Use Function CPT

Use_Func.	High	0.8302
	Low	0.1698

| Histograms for Conditional Probabilities | |

Table 12.37 Other Demands CPT

Other Dmnds	High	0.6951
	Low	0.3049

| Histograms for Conditional Probabilities | |

Table 12.38 Subjective Threat CPT

Subj_T	High	0.7382
	Low	0.2618

| Histograms for Conditional Probabilities | |

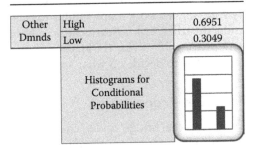

Table 12.39 Social and Economic CPT

Soc. & Econ	High	0.6238
	Med.	0.3377
	Low.	0.0385
	Histograms for Conditional Probabilities	

12.6.4 Historical (Prior) Results

We now have all the components to exercise the risk network. Figure 12.21 shows the results of the prior (historical) marginal probabilities for the particular bridge under consideration. Note that the combined historical results show a probability of high risk of 65%, while the probability of low risk is only 13%. These results are not surprising since all of the input variables that affect risk tend to be high. For example, the probability of high (demanding/less safe) site condition is 71%, while the probability of high physical vulnerability is 51%. The use of the network to process risks in this manner allows us to account for any desired interaction between variables and produces the results in a form of histograms, rather than a single value. Owners can gain more insight into risks and their underlying causes by carefully studying network maps similar to that of Figure 12.21. However, as mentioned earlier, risk assessment is only one component of the risk management process. Now that the analysis shows a relatively high security risk, the owner has to decide on what actions to take.

12.6.5 Risk Mitigation Proposition

If the high-risk result for the bridge under consideration is not acceptable, then there is a need to improve the condition. A study of the results of Figure 12.21 shows that there are two potential factors that can be improved upon: site and physical vulnerability. In general, improving other factors is possible, but they require more effort than improving site and physical vulnerability. Hence, let us assume that the owner embarked on a risk mitigation effort for the site, and the bridge physical vulnerability resulted in a probability of both low site security and low physical vulnerability of 100%. If we introduce these probabilities as evidence in the risk network, the resulting new risk network is shown in Figure 12.22. Note that the mitigation efforts have now improved the probability of low risk to 42% (up from 13%) and lowered the probability of high risk to 24% (down from 64% before the mitigation efforts). It is also worth noting that all

Table 12.40 Vulnerability CPT

Potential #		3							
Site		High				Low			
Ph_V		High	Med.	Low	High	Med.	Low		
V	High	0.9199	0.4448	0.1511	0.5353	0.0921	0.0033		
	Med.	0.0800	0.4967	0.4362	0.4292	0.5527	0.1726		
	Low	0.0001	0.0585	0.4126	0.0355	0.3553	0.8240		
	Histograms for Conditional Probabilities								

Table 12.41 Consequences CPT

	Potential #	7											
	Site	High						Low					
	Soc. & Econ.	High		Med.		Low		High		Med.		Low	
	Use_Func.	High	Low	High	Low	High	Low	High	Low	High	Low	High	Low
C	High	0.9675	0.5842	0.6999	0.2242	0.4963	0.1215	0.5353	0.3243	0.2242	0.0945	0.0385	0.0001
	Med.	0.0325	0.3758	0.2885	0.4626	0.4258	0.4148	0.4292	0.5059	0.4626	0.4493	0.3377	0.0800
	Low	0.0000	0.0400	0.0116	0.3133	0.0778	0.4637	0.0355	0.1698	0.3133	0.4562	0.6238	0.9199

Histograms for Conditional Probabilities

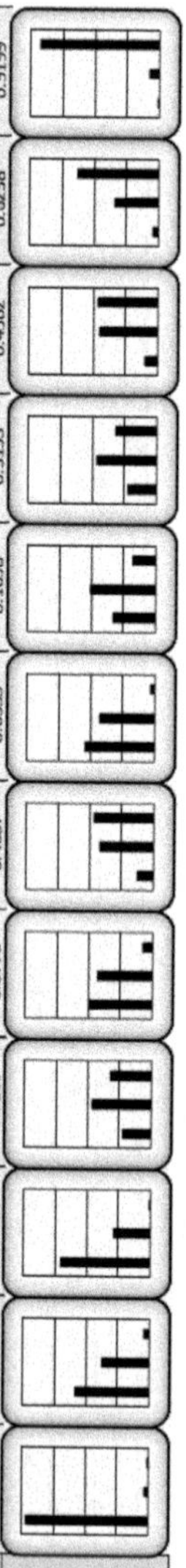

Table 12.42 Threats CPT

Potential #	8							
Site								
Subj_T	High				Low			
	High		Low		High		Low	
Other_Dmnds	High	Low	High	Low	High	Low	High	Low
T High	0.9883	0.5353	0.1215	0.4963	0.5618	0.1683	0.0963	0.0000
Med.	0.0117	0.4292	0.4148	0.4258	0.3829	0.3819	0.3894	0.0117
Low	0.0000	0.0355	0.4637	0.0778	0.0553	0.4499	0.5143	0.9883
Histograms for Conditional Probabilities								

Table 12.43 Risk CPT

Potential # 9

T = High

Ri	C	High	High	High	Med.	Med.	Med.	Low	Low	Low
	V	High	Med.	Low	High	Med.	Low	High	Med.	Low
High		0.9883	0.8683	0.6999	0.8683	0.5353	0.3216	0.3588	0.2952	0.2166
Med.		0.0117	0.1314	0.2885	0.1314	0.4292	0.4253	0.4246	0.4096	0.4246
Low		0.0000	0.0003	0.0116	0.0003	0.0355	0.2531	0.2166	0.2952	0.3588

Histograms for Conditional Probabilities

T = Med.

Ri	C	High	High	High	Med.	Med.	Med.	Low	Low	Low
	V	High	Med.	Low	High	Med.	Low	High	Med.	Low
High		0.8683	0.5783	0.1986	0.5636	0.2952	0.0754	0.2670	0.0482	0.0003
Med.		0.1314	0.4142	0.6028	0.3610	0.4096	0.3610	0.4660	0.4707	0.1314
Low		0.0003	0.0074	0.1986	0.0754	0.2952	0.5636	0.2670	0.4812	0.8683

Histograms for Conditional Probabilities

T = Low

Ri	C	High	High	High	Med.	Med.	Med.	Low	Low	Low
	V	High	Med.	Low	High	Med.	Low	High	Med.	Low
High		0.0116	0.1215	0.2166	0.2952	0.1215	0.0003	0.2166	0.0207	0.0000
Med.		0.2885	0.4148	0.4246	0.4096	0.4148	0.1314	0.4246	0.3577	0.0117
Low		0.6999	0.4637	0.3588	0.2952	0.4637	0.8683	0.3588	0.6216	0.9883

Histograms for Conditional Probabilities

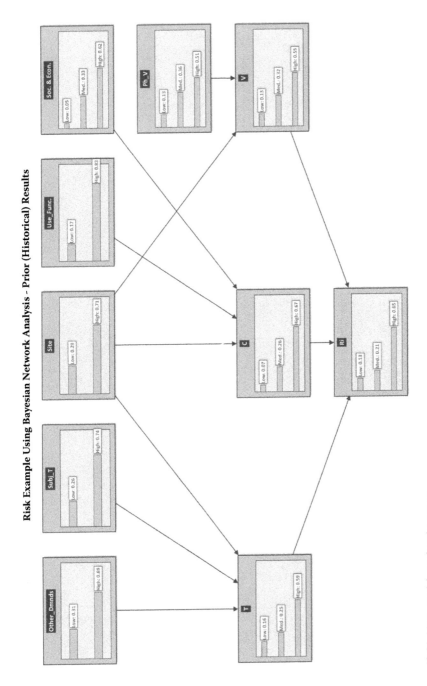

Risk Example Using Bayesian Network Analysis - Prior (Historical) Results

Figure 12.21 Historical (prior) risk picture.

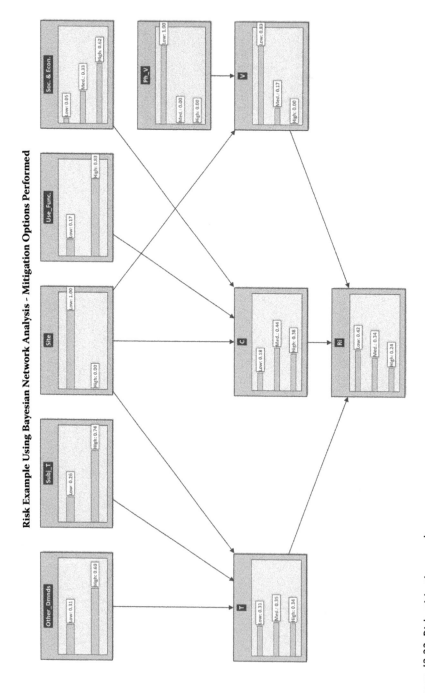

Figure 12.22 Risk mitigation results.

underlying risk components (threat, consequences, and vulnerability) have also improved. Thus, this risk network can be used to estimate the effect of various scenarios before taking mitigation action.

Although the above model is fairly simplified, it shows various steps that are needed to embark on a risk-based management. For risk assessment, the example showed the necessary components that need consideration: uncertainties, historical observations, personal judgment and experiences, intersections and links between variables, and allowances for scalability of the risk network. How the assessment network needs to be employed for developing optimal mitigation decisions is also shown above.

ACKNOWLEDGMENTS

The authors thank William Moreau for his review and comments. All the opinions and findings reported in this chapter are those of the authors and not of the organizations they represent.

REFERENCES

Agrawal, A.K., and Kawaguchi, A. (2009). *Bridge Element Deterioration Rates*. Final report on project C-01-51. Submitted to the New York State Department of Transportation, Albany, NY.

Alampalli, S., and Ettouney, M. (2008). Multihazards applications in bridge management. In *10th International Bridge and Structure Management Conference*, Buffalo, NY, pp. 356–368. Transportation Research Board Circular E-C128.

Alampalli, S., and Ettouney, M. (2010). Resiliency of bridges: A decision making tool. *Bridge Structures*, 6(1).

Alampalli, S., and Ettouney, M. (2013). Multihazards considerations for cable-bridges: A security viewpoint. Presented at Proceedings of the 8th International Cable Supported Bridge Operator's Conference (ICSBOC) 2013, Edinburgh, UK.

DHS. (2009). *National Infrastructure Protection Plan*. Department of Homeland Security, Washington, DC.

DHS. (2011a). *Integrated Rapid Visual of Screening of Buildings*. Building and Infrastructure Protection Series. Department of Homeland Security, Washington, DC.

DHS. (2011b). *Integrated Rapid Visual Screening of Mass Transit Stations*. Building and Infrastructure Protection Series. Department of Homeland Security, Washington, DC.

DHS. (2011c). *Integrated Rapid Visual Screening of Tunnels*. Building and Infrastructure Protection Series. Department of Homeland Security, Washington, DC.

Ettouney, M., and Alampalli, S. (2012a). *Infrastructure Health in Civil Engineering: Theory and Components*. CRC Press, Boca Raton, FL.

Ettouney, M., and Alampalli, S. (2012b). *Infrastructure Health in Civil Engineering: Applications and Management*. CRC Press, Boca Raton, FL.

Ettouney, M., and Alampalli, S. (2016a). *Risk Management in Civil Infrastructure*. CRC Press, Boca Raton, FL.

Ettouney, M., and Alampalli, S. (2016b). *Multihazard Considerations in Civil Infrastructure*. CRC Press, Boca Raton, FL.

Ettouney, M. Alampalli, S., and Agrawal, A. (2005). Theory of multihazards for bridge structures. *Bridge Structures*.

FEMA. (2005). *Risk Assessment: A How-To Guide to Mitigate Terrorist Attacks*. Risk Management Series, FEMA 452. Federal Emergency Management Agency, Washington, DC.

Fenton, N., and Neil, M. (2013). *Risk Assessment and Decision Analysis with Bayesian Networks*. CRC Press, Boca Raton, FL.

FHWA. (2012). *Primer for the Inspection and Strength Evaluation of Suspension Bridge Cables*. Federal Highway Administration Publication FHWA-IF-11-045. Federal Highway Administration, Washington, DC.

Gutteling, J., and Wiegman, O. (1996). *Exploring Risk Communications*. Kluwer Academic, Dordrecht, The Netherlands.

Jorion, P. (2007). *Value at Risk*. McGraw-Hill, New York.

Koller, D., and Friedman, N. (2009). *Probabilistic Graphical Models: Principles and Techniques*. MIT Press, Cambridge, MA.

Lasswell, H.D. (1948). The structure and function of communication in society. In *İletişim kuram ve araşt.rma dergisi*, Say. 24 K.ş-Bahar 2007, s.215–228 L.

Mahmoud, K.M. (2011). *BTC Method for Evaluation of Remaining Strength and Service Life of Bridge Cables*. Report C-07-11. Submitted to the New York State Department of Transportation, Albany, NY.

NCHRP. (2004). *Guidelines for Inspection and Strength Evaluation of Suspension Bridge Parallel-Wire Cables*. NCHRP Report 534. National Cooperative Highway Research Program, Transportation Research Board, Washington, DC.

Neapolitan, R. (2004). *Learning Bayesian Networks*. Prentice Hall, Upper Saddle River, NJ.

NIAC. (2009). *Critical Infrastructure Resilience Final Report and Recommendations*. National Infrastructure Advisory Council, Washington, DC.

NRC. (2010). *Review of the Department of Homeland Security's Approach to Risk Analysis*. National Research Council, National Academic Press, Washington, DC.

Vose, D. (2009). *Risk Analysis: A Quantitative Guide*. John Wiley & Sons, Hoboken, NJ.

Index

9 780367 868529